Commodities and Options Trading for Beginners

Step-by-Step Guide with Clear Examples

Elliot M. Sage

Copyright © 2024 by Elliot M. Sage

All rights reserved.

No portion of this book may be reproduced in any form without written permission from the publisher or author, except as permitted by U.S. copyright law.

Disclosure

This site is designed to provide informative and accurate facts of the subject matter covered. It is published with the understanding that the Author is not engaged in rendering legal, accounting, or other professional advice. While all attempts have been made to verify information provided on this site, the Author does not assume any responsibility for errors, omissions or contradictory interpretation of the subject matter covered.

There is a risk of loss in futures investments. Past performance is not guarantee of future results. Hypothetical or simulated performance results have certain inherent limitations. Unlike an actual performance record, simulated results do not represent actual trading. Also, since the trades have not actually been executed, the results may have under - or over compensated for the impact, if any, of certain market factors, such as lack of liquidity. Simulated trading programs in general are also subject to the fact that they are designed with the benefit of hindsight. No representation is being made that any account will achieve profits similar to those shown.

Contents

Introduction — 1

1. PART I   COMMODITIES Introduction — 5
2. Relationship Between Commodities and Futures Contracts — 9
3. The Mechanics of Trading — 18
4. Price Forecasting — 31
5. PART II - OPTIONS Introduction — 39
6. What Are Options — 47
7. Call and Put Options Explained — 52
8. Buying a Call Option — 67
9. Buying a Put Option — 80
10. Selling a Call Option — 93
11. Selling a Put Option — 105
12. Options Trading Strategies — 113

| 13. | How to Read Options Pricing Charts | 150 |
| 14. | Conclusion | 174 |

Glossary of Terms	179
About the Author	208
References	210

Introduction

This comprehensive guide is designed to guide you through the intricacies of commodities and futures, while also helping you build the practical and foundational knowledge you need to thrive in the world of options trading.

Commodities - encompass a vast array of raw materials, from agricultural products like corn and coffee to precious metals like gold and silver, each influenced by a complex network of global variables like supply and demand, weather, and geopolitics.

Futures trading – introduces the concept of contracts that obligate traders to buy or sell a specified quantity of a commodity at a predetermined price and date in the future.

Options trading - is a sophisticated financial instrument in its own right and adds another layer of complexity. Options grant the holder the right, but not the obligation, to buy or sell an underlying asset at a specific price within a set timeframe. Learning the intricacies of calls, puts, and various options strategies can be a significant challenge.

But this book is expertly crafted to simplify these complexities and make them easily accessible to any novice or beginner.

For this purpose, I have organized the book into two distinct parts:

Part I: An Introduction to Commodities and Futures Trading.

The first part of the book acts as a gateway to the world of commodities and futures trading, ensuring that you become well-versed in the fundamental concepts and practices.

Here, you'll embark on a journey of discovery, unraveling the mysteries of commodities markets, understanding the role of futures contracts, and learning how to navigate the intricacies of commodity trading.

We'll examine in detail the fascinating world of raw materials, from precious metals to agricultural products, and explore how they play an integral role in the global economy. Whether you're intrigued by energy, metals, or agricultural commodities, this section will lay the groundwork for your successful trading endeavors, helping you confidently navigate the markets, and making informed decisions to ultimately grow your wealth.

Therefore, in Part I, you can anticipate becoming acquainted with an assortment of essential definitions and terminology that you will need to grasp. This will prepare you for Part II, where the real trading and exercises begin.

Part II: Options Trading.

The second part of the book expands on the knowledge gained in the first section and delves into the world of options trading. Many people think trading options is just for stocks. But options are not limited to just stocks; they can be traded on a wide range of underlying assets, including commodities and futures.

Here, you will not only gain an in-depth understanding of options but also discover how to use commodities and futures as the underlying asset class for your options strategies.

Options can be a powerful tool for traders to manage risk, hedge positions, and create opportunities, and this section will provide you with the insights and skills needed to leverage options effectively.

Trading futures and trading options carries considerable risks. That's the first thing anyone entering the futures markets should know. But it also has very compelling profit opportunities. People have made and continue to make tidy fortunes trading commodities and using options.

Whether you aspire to be a trader, investor, or simply wish to expand your financial knowledge, this book is your stepping stone into the captivating world of commodities and options trading. Get ready to embark on a journey of learning, growth, and financial empowerment as you turn the pages of this enlightening guide.

Chapter One

PART I
COMMODITIES

Introduction

If you've ever seen the 1980's hit movie "Trading Places" you would have gotten a Hollywood style, brief introduction to the world of trading commodity futures. Commodities is the business of trading valuable necessities. The kind we use over and over again in one form or another, such as sugar, corn, wheat, soybeans, gold, silver, vegetable oils, cattle, oats and many more. The basic concept behind trading commodities is establishing the price of a product today (for example a bushel of wheat), which will then be delivered sometime in the future.

> Thus, the name ***futures trading*** *(or **futures markets**)*. This is a time-honored practice dating back to the early 19th century.

Businesses that produce commodities – as well as the businesses that use them as raw materials – need commodity futures markets to hedge against the risk of price changes.

For example, a jewelry manufacturer preparing their winter catalog half a year in advance, would like to lock in the cost of the gold they'll use by purchasing gold in the futures market for delivery in six months. Likewise, a farmer worried in the spring that by harvest time corn prices might collapse, could sell corn for delivery in September. Of course, people that trade commodities aren't all farmers and jewelers. Most are just investors that are willing to speculate on the upward or downward movement of prices for these commodities.

The point is, we'll always need commodities one way or another. Wheat to make bread. Soybeans for oil and meal. Sugar for thousands of consumption products, and on and on. These are not fad items that will suddenly disappear. Sugar has been around for hundreds of years and will continue to be around for hundreds more. Same goes for corn, oats, and just about every other commodity. This means that prices for commodities will continually fluctuate responding to the laws of supply and demand.

In essence, the commodity futures market is volatile, risky, but at the same time potentially very profitable. It is, in fact, **the richest, most profitable and liquid market in the world**. How liquid? In a single day there is more money invested in commodities exchanges than the total combined of the entire stock market! A true cash bonanza for those willing to take considerable risks.

How Commodity Futures Markets Work
Commodities trade in large units called *contracts*. For example, one contract of sugar which consists of 112,000 pounds, may be trading for 10.40 cents a pound on a particular day. As domestic and world events affect the law of supply and demand for sugar, the price for a pound of sugar will go up or down. In other words, prices for commodities, so closely tied to the laws of supply and demand, will constantly fluctuate over a short-term period.

This is one of the big advantages of trading commodities. Unlike the Stock Market, where the average return over the long run is about 10% to 15%, and where you usually hold on to your shares of stock for appreciation, in commodities your gains or losses are usually realized within a few days or weeks from the time you bought a contract. This means faster profit turnover (and potentially faster losses!).

Back to our example, 112,000 pounds of sugar x 10.40 cents a pound equals $11,648. Does this mean that you would need $11,648 to own this contract? Absolutely not! In commodities you can control any contract with a very small margin amount. This is what we call *leverage* (a powerful wealth builder just as the type you're probably familiar with in real estate). It's another big reason why so many investors are constantly turning to futures trading. With relatively small deposits you control large contracts worth a lot of money.

If you feel a bit confused, don't worry. I'll explain in more detail in the next chapter.

Chapter Two

Relationship Between Commodities and Futures Contracts

I'll begin this chapter by expanding on what a commodity is in the context of the futures markets, and why we refer to this market as futures.

> A **Cash Commodity** is an actual physical commodity that could be bought and sold. It may consist of a raw material or an agricultural product - such as wheat, corn, cattle, gold, petroleum, etc. - or could be a financial instrument, such as U.S. Treasury bonds, notes, bills, derivatives, stock index, etc.
>
> A **Futures Contract** is an obligation to buy or sell a specific quantity of a cash commodity

for a fixed price agreed upon today and to be delivered at a particular date in the future. In other words, the futures contract is based on a specific cash commodity for delivery at a stipulated future date.

The futures market exchanges where these contracts are made have been described as clearing houses and continuous auction houses for the latest information about supply and demand. They are the places where buyers and sellers connect and trade anything from agricultural products to financial instruments.

The idea is that when you agree to buy a commodity futures contract, you're essentially agreeing to buy something for a set price that a seller has not yet produced or delivered. But as you will see, simply because you're participating in the futures market doesn't mean that you will be responsible for receiving or delivering large inventories of physical commodities.

To the novice, the feverish activity on the trading floor of a commodity futures exchange market, with all the frantic shouting and signaling of bids and offers, might suggest a chaotic environment. The reality, however, is that chaos is precisely what the commodity futures markets replaced.

Since the mid-nineteen century, the commodity futures markets have served a vital economic function: provide an effective mechanism for the management of price risk. It makes it possible for those who want to manage price risk (*the hedgers*) to transfer some or all of the risk to those who are willing to accept it (*the speculators*)

Prior to the establishment of central grain markets in the mid-nineteen century, farmers carted their newly harvested crops each fall over bumpy roads, in search of buyers in the major population centers. The excess supply drove prices to giveaway levels. In addition, lack of proper storage caused grains to rot.

Come spring, shortages would frequently develop and foods made of commodities, such as corn and wheat, became luxuries. Clearly, a mechanism for organized and competitive bidding was necessary. When the first central markets were formed to meet this need, contracts were entered as for spot (immediate) delivery. Eventually, contracts were entered for forward delivery. These so-called forwards were the forerunners of present day futures contracts.

Whether it's a farmer from Kansas, an importer from Hong Kong, or a speculator from Buenos Aires, all have instant access to the markets, and all have the ability to participate in the market by having their broker buy or sell futures contracts.

The Commodity Futures Market Participants

A primary function of futures markets is hedging.

Hedgers are companies or individuals who own or are planning on owning a cash commodity, and are concerned that the cost of the commodity may change before they either buy or sell it. Individuals or entities that use the futures markets for hedging are: farmers, grain elevator operators, bankers, exporters, manufacturers, insurance companies, pension fund managers, portfolio managers, and others.

For example, a soybean processor enters an agreement to sell soybean oil to a food manufacturer. Both agree on a price today even though the oil will not be delivered for six months. Since the soybean processor does not own the actual soybeans that he will use eventually to process into oil, he is concerned that soybean prices may rise during the next six months, causing him to lose money.

To hedge against the risk of rising soybean prices, the soybean processor buys soybean futures contracts calling (at today's prices) for delivery in six months. When five and a half months have passed, the soybean processor purchases the actual soybeans in the cash market (the place where people buy and sell the physical commodities). But suppose that just as he had feared, the prices of soybeans had gone up. Well, no problem. Because he hedged in the futures market, the soybean processor can now sell his futures contracts and use the gain to offset the higher cost of soybeans.

> *The reason he can sell his futures contracts, is because as you'll learn ahead, commodities trading is a zero sum game. In other words, for every existing buy contract there is an equivalent sell contract.. If you're holding the buy contract you have to buy, if you're holding the sell contract you have to sell. So at the end of the day somebody makes money and somebody loses.*

Generally, cash and futures prices tend to move in a parallel pattern, since they react to the same factors of supply and demand.

Speculators - such as you and me - are traders that trade in the futures markets hoping to profit from price fluctuations. People that speculate play a critical role in the futures markets by providing liquidity and capital, helping to ensure price stability.

Speculators enter the commodity futures market by either buying or selling a futures contract. As opposed to the example above with the soybeans processor, a speculator rarely would have an interest in owning the physical commodity or financial instrument that underlies a futures contract. As for the speculator's decision to enter the market by either buying or selling first, it will depend strictly on each person's market expectations.

For example, you may believe that the price of oats will be going up. Thereby, you purchase one CBOT (*Chicago Board of Trades*) futures oats contract calling for you to take delivery of 5,000 bushels of oats on a specific day in May. This is known as taking a **long position**. Theoretically, it "be-longs" to you. But since you have no intention of having 5,000 bushels of oats delivered to your door step, some time before the contract expiration date you are going to sell this futures contract (hopefully at a profit). In other words, you are going to **offset** your position.

Initial Position: Buy one May oats CBOT contract at $2.00/bu

Offsetting Position: Sell one May oats CBOT contract at $2.50/bu

Net Profit (Loss) = $2,500 (5,000 bushels times $0.50)

The delivery month is important. Oats, for instance, is delivered in December, March, May, July and September. So if you bought oats that calls to be taken delivery in May, you must sell oats to be delivered in May to offset your position.

If, for instance, you are long three <u>May CBOT oats contracts</u> and you wanted to offset your position, you would sell three <u>May CBOT oats contracts</u>. In effect, all the specifications must be the same: the commodity, number of contracts, exchange and delivery month. One contract negates the other.

In the same manner, you may believe that the price of corn will be declining in the future. Therefore, you enter the market by selling a CBOT corn futures contract, in the hope of later being able to buy back identical and offsetting contracts at a lower price. In this case you are shorting the market. You're **going short** on corn.

Initial Position: Sell one July corn CBOT contract at $3.00/bu

Offsetting Position: Buy one July corn CBOT contract at $2.60/bu

Net Profit (Loss) = $2,000 (5,000 bushels times $0.40)

Even though approximately 97% of traders never deliver or take delivery of the commodity they trade, the delivery provision in a futures contract is important. Because just the fact that buyers and sellers can take delivery if they chose to, assures that futures prices accurately reflect the cash market value of the commodity at the time the contract expires.

> *If you are **long a commodity** you want it to increase in value. If you are **short the commodity**, you anticipate it declining in value. The commodity market is a **zero-based market**. At the end of each day, the books are balanced. The traders on the losing side of the market pay the traders on the winning side of the market.*

Trading Styles

Three basic trading styles:

The **Scalper,** who trades in and out of the market many times throughout the day in hopes of taking small profits on heavy volume of trading. He rarely holds a position overnight. Most floor traders are scalpers.

The **Day Trader**, who is similar to the scalper in that he doesn't hold overnight positions. However, he makes fewer trades and hopes to profit from intraday trends.

The **Position Trader**, maintains a position for days, weeks or even months. The position trader is more focused on major long-term price trends.

Chapter Three

The Mechanics of Trading

Before we discuss the trading process, let's review these key commodities futures trading terms you'll need to be aware of.

Volume and Open Interest

Volume and Open Interest are two frequently cited terms you will hear in futures markets terminology. They help to gauge the level of commercial activity in a market.

Each unit of **volume** represents a contract traded, and that includes the long and the short side of the trade. It's typically quoted on a daily basis.

On the other hand, **open interest** refers to the number of futures positions that remain open, or unliquidated, at the close of each trading session.

For instance, let's assume that you buy 10 contracts and sell 7 of them back to the market before the end of the trading day. Your trades add 17 contracts to that day's total volume figure. But since you're still holding 3 contracts which remain unliquidated, open interest in the market have increased by 3 contracts to the day's total open interest figure.

The Tick

A tick is the smallest increment of change allowable in the price of a commodity. The term is a throwback from to the days when ticker tapes were used to communicate prices. Commodity markets have different minimum tick sizes.

For instance, grains have a minimum tick size of 1/4 of 1 cent. This translates into a $12.50 move on CBOT contracts ($0.0025 times 5,000 bushels). Therefore, if you're quoted a two-tick move, this means a $25.00 move on this contract.

In addition, tick sizes, as well as contract sizes, may vary with each exchange. For example, the Mid American Exchange (MidAm) trades corn in 1000 bushel contracts, with a tick size of 1/8 of 1 cent. So one MidAm tick move represents $1.25 ($0.00125 times 1000 bushels).

Interpreting Price Quotations

The different commodity contracts have different delivery abbreviations. You must familiarize yourself with these one-letter month codes in order to accurately identify the price.

One-Letter Month Codes

MONTH	CURRENT YEAR	FOLLOWING YEAR
JAN	F	D
FEB	G	E
MAR	H	I
APR	J	L
MAY	K	O
JUN	M	P
JUL	N	T
AUG	Q	R
SEP	U	B
OCT	V	C
NOV	X	W
DEC	Z	Y

Futures Abbreviations

Meat
LC	Live Cattle	Cents per pound
FC	Feeder Cattle	Cents per pound
LH	Live Hogs	Cents per pound
PB	Pork Bellies	Cents per pound

Grains
W	Wheat	Dollars per bushel
C	Corn	Dollars per bushel
O	Oats	Dollars per bushel
S	Soybeans	Dollars per bushel
SM	Soybean Meal	Dollar per ton
SO	Soybean Oil	Dollar per pound

Metals
SL	Silver	Dollars per ounce
GC	Gold	Dollars per ounce
HG	Copper	Cents per pound
PL	Platinum	Cents per pound
PA	Palladium	Cents per pound
AL	Aluminum	Cents per pound

Energy
HO	NY Heating Oil	Cents p/gallon
HV	NY Unleaded Gas	Cents p/gallon
CL	Light Crude Petroleum	Dollars per barrel

Food/Fiber
LB	Lumber	Dollars per 1000 bd. ft.
KC	Coffee	Cents per pound
CC	Cocoa	Dollars per ton
SB	Sugar	Cents per pound
CT	Cotton	Cents per pound
OJ	Orange Juice	Cents per pound

With this information, you should be in a good position to read most any quote system. For example, the following is a typical stream of quotes from a price quotation service, containing three different quotes for three different commodity markets:

LC 6658 Q   C 190 N   GC 3849 Y

This is telling us that:

1) August Live Cattle is currently trading at 66.58 cents a pound (**LC 6658 Q**).

2) July Corn is trading at $1.90 per bushel (**C 190 N**).

3) Next year's December Gold is currently trading at $384.90 per ounce **(GC 3849 Y)**.

Reading Commodities Futures Prices

Whether online or on the business section of national and local newspapers, you'll find published futures prices with volume and open interests. For example, the following chart is for wheat traded at the CBOT:

- *Wheat (CBOT) - 5,000 bu minimum - dollars per bushel*

- *Estimated volume 11,000   16,950 open interests*

- *Wheat (CBOT) - 5,000 bu minimum - dollars per bushel*

- *Estimated volume 11,000   16,950 open interests*

SEASON			OPEN INTEREST	OPEN	HIGH	LOW	SETTLE
HIGH	LOW						
6.32	3.74	SEP	5,174	4.45	4.48	4.42	4.465
6.3250	3.63	DEC	42,616	4.47	4.47	4.40	4.45
6.18	3.95	MAR	10,630	4.45	4.46	4.42	4.43
5.47	4.16	MAY	579	3.94	3.99	3.93	3.95
4.65	3.70	JUL	3,938	3.93	3.94	3.93	3.935
4.10	3.04	SEP	43	4.01	4.01	3.97	4.00
4.66	3.98	DEC	84	3.94	3.98	3.93	3.97
3.96	3.65	JUL	44	3.90	3.94	3.90	3.92

Explanation of the above chart is as follows:

Season High/Low = These two columns display the highest and lowest prices ever reached for each delivery month since the contract began trading.

Open Interest = The total outstanding positions for each contract month.

Open = The opening price at which the first bids and offers were made and the first transactions were completed.

High = The top price at which a contract was traded during the trading day.

Low = The lowest price at which a contract was traded during the trading day.

Settle = The official daily closing price. The basics of trading commodities begins of course with deciding on what to trade. Once you make the decision, you place an order by contacting a broker. You instruct your broker on the type of order you're placing. The broker then time stamps the order and delivers it to the trading desk of the floor, or the trading pit. When your order is matched, it is confirmed to your broker and then you. Trading pits are each divided into sections designated for trading commodities in particular contract months. No trading is allowed outside a contract's assigned pit.

There are many ways to enter orders into the futures markets. Following are the most commonly used orders.

Market Order

This is the most common type of order. You simply state the number of contracts you want to buy or sell for a particular delivery month. You don't specify price, since your objective is to have your order executed as soon as possible.

Limit Order

A limit order specifies a price limit at which the order must be executed. The advantage is that you know the worst price you'll get if the order is executed. The disadvantage is that you can't be certain that the order will be filled.

Day Order

An order that automatically expires if it is not executed on the day it is entered.

Stop Order

Stop orders (also known as stop-loss or sell-stop orders) are not executed until the market reaches a certain price, at which point they become market orders. They are normally used to liquidate earlier positions. They can also be used to enter the market. For instance, suppose you expect a bull market, but only if the price passes through a specified level. In this case, you could enter a buy-stop order to be executed if the market reached that point.

Position And Price Limits

To ensure orderly markets, futures exchanges set both price and position limits. A position limit will restrict the number of contracts that may be held by a market participant. Price limits, also called daily trading limits, specify the maximum price range allowed for each contract.

Margins

Margin is the amount of money deposited by both buyers and sellers of futures contracts to ensure the performance of the terms of the contract. In futures, the margin is not a payment of equity or a down payment as in securities, but rather a performance bond or security deposit.

There are two kinds of margins. The initial amount deposited into your account when you first place an order is called the **initial margin.** This amount is debited or credited daily on the close of the day's trading session. This is called "*mark to the market*."

The second kind of margin is called a **maintenance margin.** It is usually a little less than the initial margin, and it is the minimum amount you must maintain in your account after a position has been opened. For example, let's say you enter the market by placing an order to buy. The initial margin required for your particular contract is $400. Once you hold the position, the maintenance margin may drop to $300. This means you need to show $300 in equity in your trading account for each contract you acquire.

If the debits from a market loss reduce the funds in

your account below this maintenance level, you will be required to restore the account to the initial margin level. This request for additional money is called the **margin call.**

For example, you have $1,000 in your account. You purchase one corn contract requiring $550 initial margin, and your maintenance margin requirement is $500. You paid $2.50 per bushel. Therefore, you control a $12,500 contract ($2.50 times 5,000 bushels). If corn prices drop to $2.46, your account would be debited $200 ($2.46 times 5,000 bushels = $12,300). Your equity is now $800. If corn continues to drop to $2.40, your account will be debited $500 ($2.40 times 5,000 bushels = $12,000). Your equity is now $500, which is the minimum maintenance margin. If corn moves below $2.40, for instance to $2.38, you will receive a margin call for $100. This would bring your equity back to the minimum maintenance level of $500.

The exchanges firmly enforce margin calls to protect the integrity of the market. In the same manner, margin in excess of the required amount is available for withdrawal daily.

Commodities Trading Exchanges

Not all commodities trade in the same exchanges. Following is the breakdown of the main exchanges in the United States:

1. **Chicago Board of Trade (CBOT)** : corn, wheat, soybeans, soybean oil, soybean meal, oats, gold, notes, T-bonds, municipal bonds, 30-day Fed funds.

2. **Chicago Mercantile Exchange (CME):** cattle, hogs, pork bellies, gold, silver coins, lumber, Eurodollars, certificates of deposits, currencies, Treasury bills, S&P 500 Index
futures.

3. **Coffee, Sugar & Cocoa Exchange (NYCSCE):** coffee, sugar, cocoa.

4. **Commodity Exchange, Inc. NY (COMEX)**: copper, silver, gold.

5. **Kansas City Board of Trade (KCBT):** wheat, Value Line Stock Index futures.

6. **Mid America Commodity Exchange (MidAm):** live cattle, hogs, oats, rough rice, soybeans, wheat, corn, gold, silver.

7. **Minneapolis Board of Trade (MPLS)**: wheat, sunflower seeds.

8. **New York Cotton Exchange (NYCTE)**: cotton, orange juice, propane gas.

9. **New York Futures Exchange (NYSE):** New York Stock Exchange Index, option on futures.

10. **New York Mercantile Exchange (NYMEX):** crude oil, unleaded gasoline, heating oil, natural gas, palladium, platinum, propane.

Chapter Four

Price Forecasting

If commodities prices were forecastable with accuracy and precision, it wouldn't be tradable for a profit, since everyone would know the answer in advance. But with help from a variety of forecasting techniques, we can at least come up with an educated guess on the direction of prices.

The Fundamental Market Analysis

The fundamental market analysis is based on supply and demand information. That is, if you expect increased demand for a product, or a scarcity of supplies for a product, then prices should rise. Conversely, decreased demand or excess supplies should drive prices down.

In principle, fundamental analysis is actually simple. The theory is: you take whatever is left over from last year (carryover), and add it to this year's production (supply), then subtract this year's usage. This should give you a good estimate of available stock levels, from which you can draw price projections.

So, fundamental theory =

(Carryover + Current Year Production) MINUS **Last Year's Usage** EQUALS = **Available Stocks**

The reality, however, could be much more complicated. To begin with, carryover may be stored in places where it is extremely difficult to measure. Such as farms in remote parts of the world or uncooperative countries. Also, some carryover stock might have suffered from spoilage, rendering the stock unusable. Additionally, there may be insurmountable transportation problems to move the stock to where it's needed.

Production can be just as troublesome. Weather imponderables will have an unpredictable impact on yields. And demand is very tricky to forecast, to say the least.

Nevertheless, the fundamental analysis is the most

helpful method we have to determine price forecasting and long-term market trends. Much of the fundamental trade centers upon the release of key agricultural and financial reports.

The following are some of the major United States Department of Agriculture (USDA) reports:

REPORT   SCHEDULE

Grain Stocks ....January and end of March

Prospective Plantings.......End of March

Crop Production.............Monthly, April through Dec,

Crop Progress ....Weekly, April through Dec.

Cattle on Feed..................Monthly

These reports concentrate on a series of *supply* and *demand* factors that may be summarized as follows:

The Supply Factors

PLANTING PROJECTIONS: most agricultural markets follow a yearly crop cycle, beginning with plantings and concluding with harvest. At the start of the season, the market assesses the supply outlook by estimating acreage expected to be planted in each crop.

CARRYOVER INVENTORIES: if carryover supplies are considered excessive, prices will be held down.

WEATHER/CROP PROGRESS: throughout the growing season, crop projections begin to focus on crop yields. Weather becomes a major factor. As growing conditions become better or worse, prices adjust accordingly.

INTERNATIONAL COMPETITORS: the progress of major foreign crops are closely monitored to assess the supply outlook on a global scale.

GOVERNMENT PROGRAMS: domestic support programs may increase or decrease the amount of acres planted in various crops. Also, export policies and international trade agreements may impact prices.

FOREIGN GOVERNMENT PROGRAMS: just as the United States, other countries may subsidized their exports, bringing additional supplies into the world market.

The Demand Factors

LIVESTOCK REPORTS: livestock inventories may impact prices, since cattle, hogs and chickens consume great amounts of corn and soybean meal. As a result, an increase in the number of cattle on feed, would increase the demand for corn and drive corn prices higher.

CONSUMER PREFERENCES: as consumers' food preferences change, the demand for certain crops will be affected. For example, when studies indicated that oats were beneficial for cutting down cholesterol levels, the demand for oats increased leading to higher oat prices for a time.

CURRENCY MARKETS: as the value of the US dollar fluctuates, the relative cost of US agricultural products in relation to foreign agricultural products will change. For example, a very strong dollar may hurt the demand for US crops.

FOREIGN PURCHASES: major countries, such as China and former Soviet Union, are continuously assessed for their potential import demand.

The Technical Analysis

The technical analysis is another method of forecasting prices. The technical trader is a pure technician that is not concerned - as the fundamentalist may be- with understanding why the market moved the way it did. Instead, he attempts to predict market price direction by detecting patterns of price behavior that have signaled major movements in the past.

Technical traders rely on bar charts to identify price trends, special patterns, trend formations, and areas of support and resistance. Price support occurs when there is sufficient buying of the futures contract to oppose a price decline. On the other hand, where a market rallies time and again only to stall out at a certain price level, is known as a resistance area.

You will find text books of encyclopedic dimensions, discussing the interpretation of bar charts for commodity futures trading. Following, is a brief synopsis of the most common price patterns and trend formations:

UPTREND: this is a sequence of higher highs and higher lows. By connecting the low end of the prices you draw a trendline at about a 45 degrees angle. The closer to 45 degrees, the more accurate the trend is believed to be. A major uptrend is usually accompanied by increases in volume and open interest.

DOWNTREND: this is a sequence of lower highs and lower lows. In this case, the trendline is drawn along the top of the prices. Again, the closer to 45 degrees the better. Just like the uptrend, a major downtrend will usually show increased volume and open interest.

TOP: indicates a probable end to an uptrend. A double top is even a stronger indicator that a uptrend has ended.

BOTTOM: indicates the probable end of a downtrend. Again, a double bottom would even be a stronger indicator.

HEAD AND SHOULDERS (TOP): signals a major reversal from an uptrend to a downtrend. It's detected by the formation of a four phase pattern. That is, the formation of a left shoulder, the head, the right shoulder, and the penetration of the neckline.

HEAD AND SHOULDERS (BOTTOM): indicates a major reversal from downtrend to uptrend. It's spotted by the same four phase formation, only inverted.

TRIANGLES: there are three types of triangle patterns - the ascending, the descending and the symmetrical. The ascending triangle indicates a breakout of prices on the upside. The descending points to a breakout on the downside. And the symmetrical triangle signals that a major move out of a consolidation phase will occur, but does not uncover the direction of the move.

Technical traders continuously monitor patterns such as these to discern price movements. For example, a head and shoulders top pattern is taking shape, and the trader recognizes the formation of the left shoulder and the head phase. He may choose to go short in anticipation of a major decline in prices.

Critics of the technical analysis method of trading, suggest that technical analysis is a self-fulfilling prophecy, stemming from the fact that it is so widely accepted by its followers. A technical analysis is built on the premise that history repeats itself, and technical traders expect it to continue. They are conditioned to believe that when a major trendline is broken, it's time to jump off and reverse their position. And with a clock's precision they will all reverse their positions, thereby fulfilling its prophecy.

Chapter Five

PART II - OPTIONS

Introduction

In this Part II of the book, I will introduce you to the fascinating world of options using commodities and futures as our trading asset (**though the concepts and dynamics of options trading apply just as well to the stock market**). My intent is to lay down a solid foundation for you to build upon, by describing in a clear and simple fashion what options are all about. By the end of this book, you'll understand the mechanics of options, uncovering the incredible opportunities for financial gains, while learning to assess the potential risks.

Everyday, people across the country miss out on fantastic profit opportunities. Why? Simply because of fear and misunderstanding. Fear is like a ten-foot giant standing in your way, and keeping you from reaching personal accomplishments and worthwhile financial objectives. When you don't know how something works, you can't help feeling anxious and uncomfortable about it. The best weapon against fear is knowledge.

You have an interest in options, and you've recognized that the first thing you need to do is learn more about them. So before we go on, let me congratulate you on this important step you've taken to obtain that knowledge.

When it comes to options, not too many people have the knowledge it takes to trade them. A lot of people prefer investing in much more popular markets such as real estate, or buying shares of stock through the stock market, simply because they're familiar with these types of investments (or at least they think they are.) But outside of this mainstream, there's a smaller part of the population who does understand the nuts and bolts of options trading, and know exactly well the exciting financial opportunities they offer. And while there are always elements of risk to consider with options, the knowledgeable investor will be in a better position to substantially limit these risks.

Overall, options are relatively new. They began trading publicly in 1973 in the stock market, and only since the early 1980s on regulated futures exchanges. As an investment, they could be considered highly speculative or highly conservative. It all depends on your objectives.

Why do People Trade Options?

People trade options for different reasons. But most of all it's because of its build-in flexibility.

> On the *speculative side*, you may be buying or selling options in anticipation of a market price move you think will go your way. You may stand to earn a large profit with one option. Or, it's possible that even with a few profitable options you could more than make up for losses incurred on other options that didn't quite go your way.

> In contrast, on the *conservative side*, you may be trading options as an 'insurance' or protection against a possible loss from a futures contract you're holding. In this case, you're using it to hedge a risk position.

> And in between these two extremes of speculation and conservatism, there's a whole range of possibilities that exist for trading options. They're all based on different strategies and techniques you may employ, designed

to increase your profit potential and/or limit your risk exposure.

For example, as you'll soon learn in the following chapters, there are *call option*s and *put options,* and you can either *buy* them or *sell* them. Therefore, options allow you to profit no matter what your outlook of the market at the time may be. You might trade options because you're bullish about the market, but want to limit your risk, or you're bearish and want to limit your risk. Or you might trade options because you think there's going to be a large price move, upward or downward. Then again, you might even trade options because you believe that there won't be a price move at all! All of these scenarios are possible, and any one of them can make you money.

So as you can see, trading options offers enormous flexibility. This is one of their main attractions. They are, by far, much more flexible directly buying stocks or futures contracts could ever be.

In addition to their flexibility, options - whether you're buying or selling them - have some other extremely appealing aspects:
a) They offer the option buyer the potential for earning unlimited profits while limiting the risks to only the up-front cost of the option (known as a premium) plus

transaction and commissions costs;

b) They offer the seller of an option extraordinary probabilities of profits by putting the odds heavily in their favor;

c) Last but not least, they offer the option trader an incredible amount of leverage. As the owner of an option on a certain 'asset', you control that asset - which is worth a great deal of money - for only a fraction of its cost.

In other words, you put up a premium, which is a small amount of money in relation to the value of the asset you're going to control. In turn, this asset has unlimited profit potential. What this means to you, is that because you have so little capital tied up, your returns on this investment are enormous. Also, since you don't need to tie up a lot of money to buy the asset itself, you could instead put this money to work in some other investment. And on top of it all, you can limit your risk to just the little money that you tied up. Indeed, this is what you call a pre-defined risk.

Complex Subject Made Easy

The subject of options can be fairly complex. My greatest challenge in this book is to communicate these complicated concepts by explaining them in the simplest way so it makes sense to everyone. But most importantly, by explaining things in a way that will give you immediate useful application. If this book were about automobiles, I would have started it by first explaining the benefits of owning one. Then, I'd go on to explain some basic parts of the automobile, such as the steering wheel, the breaks, review mirror, etc., giving you examples of how you would use them. Then I'd take it one step further, and begin to describe some of the more technical aspects of an automobile, such as the internal combustion process, and give you a more sophisticated understanding of what keeps a car running. Finally, I would show you what you've been waiting for all along, which is how to drive and get the best use out of your car.

You'll find this book to be very similar to this automobile example. I'll begin by introducing you to the jargon of options and how it all works together. And continually be reinforcing these concepts with simple examples you can follow. As you become comfortable with options theory and the language of options, I'll cover in greater detail the mechanics of trading. Also show you how to read and interpret option prices for all of the main futures markets. But I feel that, by far, what you'll enjoy the most from this book, will be learning the best strategies and combined techniques used for trading. These techniques are designed not only to make you money, but also to make it the safest possible way.

Even though I won't bog you down with excruciating details, I guarantee that by the time you're done reading this book, you're going to know more about options than most people out there that currently trade them. Is this because I'm about to reveal trading secrets unbeknownst to anyone but me? No, of course not. It's simply because so many people are in such a hurry to 'make' money, that they don't bother to 'learn' what they need to know to make that money safely. They 'heard' trading options was a good deal and they're ready to jump on it.

In fact, let me make a comparison using our previous automobile example. Imagine an anxious teenager who just got his driver's license and got a hold of dad's brand new car. Do you think the average teen has the patience to seriously listen to you about regular tune-ups and oil changes as he speeds out of the driveway?

There are people in a hurry to get rich and that's just the way it is. And if that means going down a blind alley, so be it. That's what they'll do. However, we won't go down this alley blind. Instead, we'll be building a strong foundation. Those who don't have this type of foundation depend exclusively on their brokers. But let's face it, even though there are some excellent brokers out there, they do have to make a living. And that living comes off your commissions, so you can't always count on that their recommendations being entirely in your best interest.

Chapter Six

What Are Options

Let's begin by understanding exactly whar options are:

> Options are the right - but not the obligation - to buy or sell a futures contract (or any other asset) at a specified price at any time prior to a specified date.

Let's say that it is now the month of July, and that you've just learned that your neighbor wants to sell his fantastic twin-engine snow blower that he's got stored away in his garage. He wants $1,000 for it, and would like to have it sold by January 15th. You'd like to have it, but unfortunately you don't have the money to buy it right now. Yet, you're convinced it's going to be a snowy winter and the snow blower is going to be worth a lot more than $1,000 by January. So you and your neighbor agree on an OPTION. Whereby, you have the RIGHT (but not the obligation) to BUY his snow blower for the

SPECIFIED amount of $1,000, on or before the SPECIFIED date of January 15th. However, to grant you this right, your neighbor required that you put up front $100. And of course, as the seller of the option, your neighbor gets to keep the $100 even if you ultimately decide to pass on the snow blower.

> The above simple-minded illustration, manages to cover some of the important components of an option which we'll be dealing with throughout this guide. We have the **underlying asset** (the snow blower), the **strike price** ($1,000), the **premium** ($100), and the **expiration date** (January 15th.).

Let's look at another example, only this time using real estate and with a slightly different angle. You'll notice that the same basic option components will all still be present throughout this next example.

Say there's a vacant lot in your neighborhood worth about $100,000. Because of improvements in the area, you have reason to believe that this same lot will be worth much more than $100,000 in one year. So you go to the owner of the vacant lot and offer her an OPTION. You want the RIGHT (but not the obligation) to buy her land anytime between now and the next 12 months, for the SPECIFIED price of $115,000. In order to grant you this option, she requires a PREMIUM (up front deposit) of $8,000. She'll keep the $8,000 whether you buy the land or not.

Why does the landowner require this premium? Simply because it's the landowner that carries the greatest financial risk during the next 12 months. Suppose a property investment company comes along and wants to build a mall right next to this vacant lot, and would like to use the lot for additional parking. They offer the lot owner $300,000 for it. However, she's not free to sell it to them over the next 12 months, because she has the obligation to sell it to you for $115,000 anytime over the next 12 months. When the investment company contacts you with the offer, you EXERCISE your RIGHT to buy the lot at the SPECIFIED price of $115,000, and then turn around and sell it to the property investment company for $300,000. In other words, you've just made $185,000 profit on an $8,000 investment.

On the other hand, what would happen if the opposite occurs? Suppose that instead of a mall they build a lovely toxic waste dump right next to the land. The value of the lot will probably decrease. The owner comes to you and asks if you would like to exercise your right to buy the lot now. You answer "I might be slow, but I'm not stupid. I do not wish to exercise my option!" So after the 12-month period, you lose the $8,000 premium you put up for the option, but at least you're not stuck with a worthless piece of land.

At this point, you're probably figuring that as an option buyer you stand to make a lot of money when things go your way, and only lose a little money when things don't go your way. This is true. The question is, however, how often will things go your way and how often will they not.

In the same manner, you're probably figuring that as an option seller, even though you keep the premium, you potentially stand to lose a lot of money if the asset you own greatly increases in value. This is also true. However, the key here is what are the odds that your asset will increase or decrease so substantially in value within a defined period of time.

Think about it for a moment, how many times would your vacant lot increase three times its value because it happens to be next to a future mall they're building? Or tumble in value because it's next to a future toxic waste dump? It's not an everyday event is it? Chances are, over the next 12 months this vacant lot would have increased or decreased in value only a reasonable amount. Perhaps not even enough for the option buyer to exercise his right.

In other words, the option seller from our previous real

estate example just made $8,000 for simply granting you the right to do something that was unlikely to happen. This is why it is often said that selling options, while carrying more risks, allows you to create a stream of income.

Chapter Seven

Call and Put Options Explained

Options come in two varieties: **Call Options** and **Put Options.**

> **Call Option**
> A **call option** buyer acquires the right (but not the obligation) to buy a particular futures contract at a specified price at any time during the life of the option (usually a few months.) A buyer of a call option hopes to profit from an increase in the futures price of the underlying commodity (*also called 'underlying instrument', 'underlying contract' or 'underlying asset'*)

For example, you believe the price of gold is going up. So you buy a March 500 gold call option, paying a premium of $1,100. This option gives you the right to buy a 100 ounce gold futures contract (the underlying commodity) for $500 an ounce. Suppose that sometime before the expiration of the option the price of gold goes up just as you had anticipated. So gold is now selling for $540 an ounce. As the owner of the call option you can either exercise the option, meaning that you can buy the underlying commodity (the 100 ounce gold contract) at the stipulated price of $500 an ounce, and then sell back the futures gold contract in the open market for its current market price of $540 an ounce. Or, you can avoid all that trouble, and simply sell back the option at the current price of $540 an ounce, representing a gain of $40 an ounce. On 100 ounces your total gain is $40 x 100 = $4,000. Minus the $1,100 premium, your net profit is $2,900 (Broker's commission not factored in).

The person who sells the option receives in return the nonrefundable payment known as the premium (The seller of an option is said to be in a **short position**. The buyer of an option is said to assume a **long position**).

> **Put Option**
> A put option buyer acquires the right (but not the obligation) to sell a particular futures contract at a specified price at any time during the life of the option (usually a few months.) A buyer of a put option hopes to profit from a

decrease in the futures price of the underlying commodity. Exactly the opposite of what you want to happen when you buy a call.

For example, you think the price of gold is going down. So you buy a March 500 gold put option, paying a premium of $1,100. The option gives you the right to sell a 100 ounce gold futures contract (the underlying commodity) for $500 an ounce. As you expected, the price of gold dropped to $440 an ounce. Being the owner of the put option, you can either exercise the option, meaning that you can sell the underlying commodity (the 100 ounce gold contract) at the stipulated price of $500 an ounce,4 and then buy back the gold contract in the open market for $440 an ounce, pocketing the $60 an ounce difference. Or you can simply sell back the option at the stipulated price of $500 an ounce and realize a gain of $60 an ounce. On 100 ounces your total gain is $60 x 100 = $6,000. Minus the $1,100 premium, your net profit is $4,900.

A point to clarify is that you can sell a commodity even if you don't own it. This is explained later in '**Offsetting Transactions**'.

Strike Price
This is the exercise price. In other words, it's the specified price in which the buyer of the

option has the right to purchase a specific futures contract, or at which the buyer of a put has the right to sell a specific futures contract. In our gold futures contract example, the striking price was $500 an ounce in both cases. Expressed simply as 500.

Premium
This is the money that an option buyer is requested to pay, or that an option seller demands to receive at the time of the transaction. In other words, it's the price of the option. In our gold example, it was $1,100. You'll normally find this amount expressed without dollar signs. So when you're quoted an option at 5, it means the premium is $500. In our gold example, a broker would have stated that the *"option was at 11"*.

Essentially, premiums are arrived at through open competition between buyers and sellers on the trading floor of the exchanges, in the same way that prices of commodities are arrived at. This means that the whole process of establishing a premium on an option is strictly a function of supply and demand. When the price of a commodity is rising, the demand for call options rises accordingly, and the premiums set for that particular futures contract will be higher. In contrast, when the price of a commodity is falling, the demand for put options rises accordingly, and the premiums set for that particular futures contract will be higher.

So you can see why the premium price can vary month to month. While an option might be worth a premium of $1,100 one month, it could drop to $200 or $300 the next month. Or it could jump to $2,000 or $3,000. The two components that contribute directly in the value of a premium are: time value and intrinsic value. In short:

> PRICE OF AN OPTION = PREMIUM AND PREMIUM = INTRINSIC VALUE + TIME VALUE

Intrinsic Value & Time Value

An option's premium is made up of two parts. Its **intrinsic value** is the part of the premium that is *in the money.*

> ***In the money:*** when the market value of the underlying asset is higher than the call's striking price, or lower than the put's striking price. It means the option is already worthwhile to exercise and commands a higher premium. For example, you buy a gold call option and pay a premium of 3 points ($300). The strike

price is 500 at the time when gold is going for $400 an ounce. Or, you buy a gold put option and pay a premium of 3 points ($300). The strike price is 500 at the time when gold is going for $600 an ounce.

Its **time value** is any difference after that. In other words, time value is the amount by which the option's premium is above the option's intrinsic value. Time value reflects any extra amount that a buyer is willing to pay for the option, (or the seller is willing to accept) hoping that any changes in the underlying contract's price during the life of the option will increase its intrinsic value. Therefore, the premium of an option that's *at the money* or *out of the money* is entirely a reflection of its time value.

> **At the money**: when the market value of the underlying asset is the same as the striking price of the option. In this case, the option has no intrinsic value. For example, you buy a gold call option and pay a premium of 2 points ($200). The strike price is 500 at the time when gold is going for $500 an ounce.
>
> **Out of the money**: when the market value of an underlying asset is lower than the call's striking price or higher than the put's striking

price (the exact opposite of in the money.) In this case, the option has no intrinsic value. For example, you buy a gold call option and pay a premium of 2 points ($200). The strike price is 500 at the time when gold is going for $400 an ounce. Or, you buy a gold put option and pay a premium of 2 points ($200). The strike price is 500 at the time when gold is going for $600 an ounce.

Time value declines over the life of the option. So the time value of an option is worth less and less as the expiration date approaches, because there's less time remaining for the option to develop intrinsic value. At expiration time, the time value of an option is zero. This is why an option is said to be a **wasting asset.** Consequently, the farther away from the expiration date of an option, the higher the time value of the option.

Example: *It's January and you buy a May 650 soybeans call option (650 is the strike price and it means $6.50 per bushel). The premium for this call option is 10 cents, in other words, $500.. (Note: Soybeans trade in contracts of 5,000 bushels. So 10 cents x 5,000 bushels = $500.)* At the time you purchased the option, soybeans were trading for $6.50 a bushel. This means you were at the money, so the entire value of the option's premium represents time value. This means you're putting up the $500 premium just because of your perception of the market, and your expectation that the underlying commodity - in this case soybeans - will go up in price. You have no intrinsic value right now:

$6.50 market price of soybeans – $6.50 strike price of option = $0 intrinsic value

Suppose that during the life of the option, the price of soybeans goes up to $7.20 a bushel. Now you're in the money, and your option has intrinsic value:

$7.20 market price of soybeans – $6.50 strike price of option = $.70 intrinsic value

If you decide to exercise your option right now you'll make:

.70 x 5,000 bushels = $3,500
less premium – $500
Profit = $3,000 (minus commission)

What's more, those who want to get in on this option are going to have to pay a higher premium for it.

Now suppose instead that the price of soybeans dropped to $6.25 a bushel. Your option is now out of the money. Again, you've got no intrinsic value at this point:

$6.25 market price of soybeans – $6.50 strike price of option = ($0.25) intrinsic value

In essence, when you buy an option and you pay a

premium, you can easily recognize how much of that premium is time value and how much of it is intrinsic value. Simply compare the current market value of the underlying commodity with the option's strike price.

NOTE: An option that is deep out of the money (when there's a more substantial difference between striking price and market price), will have less time value than an option that is only slightly out-of-the-money. Simply because it would be more difficult for the deep out of the money option to become profitable.

Exercising an Option

To exercise the option means that you elected to take ownership of the underlying asset. In other words, if you exercise a call option you are purchasing the underlying futures contract at the option's strike price. Thus, you've acquired a long position on that futures contract. Likewise, if you exercise a put option you are selling the underlying futures contract at the option's strike price. You would then be acquiring a short position on the underlying futures contract.

Expiration Date

This is the last date on which the option can be either exercised or offset. One important feature here, is that the expiration of a futures option is **one month before** the expiration of the futures contract. The month that is stated on the futures option actually corresponds to the futures contract. In other words, using our previous example, we purchased a 650 May soybeans call option. The month of May actually stands for the underlying soybeans futures contract expiration. But the option itself expires in April. It's always a good idea to check the exact expiration date of the option with your broker. And remember, once the option expires you lose your premium.

Offsetting an Option

Anytime prior the expiration of the option, you can **liquidate** the option. In other words, when you sell back an option you had already bought, you're offsetting the option. This is called an *offsetting transaction*.

In the same manner, you can short sell an option. How do you sell something you don't own? Well, you borrow it first, and then you sell it. Just like when you borrow a book from the public library, you're expected to bring it back. When you sell an option through the futures exchange, you'll have a 'pending offsetting transaction' waiting. It'll be satisfied when you buy back the option through an offsetting transaction in that same exchange.

As a practical matter, you, I, and almost everyone else would offset their option. Very few options are ever exercised. Most options are offset. What is bought is usually sold back, and what is sold short is usually repurchased.

But what happens if you couldn't buy or sell back your option? Well, then you would have to exercise the option. However, this is extremely unlikely! One of the most attractive features of the options market is that it is **very liquid**. Options trade in major exchanges worldwide. Traders are buying and selling all the time. Being practically assured that you'll be able to offset an opening transaction via a closing transaction is one of the benefits of trading options.

Commissions

Commissions vary from broker to broker. But unlike commissions on futures contracts, commissions on options usually aren't round-turn (*round-turn means it covers both the opening transaction and the closing transaction*). You're charged a commission when you buy an option and when you sell an option. So you need to check with your broker about his commission schedule on options. Also, you'll be charged the normal fees that you're charged in a futures contract. That is, NFA and CTFC fees, exchange fees, etc. Varies but usually less than $100.

Here are some additional terms you might come across:

Volatility

This is a measure of the degree or likelihood of price change in a commodity (or stock, or index), during an X amount of time (usually 12 months.) The higher the volatility, the greater the chance of the commodity moving in price (*in either direction, since volatility is not direction-biased.*) Conversely, the lower the volatility of a commodity (or stock, or index), the less chance of a price move.

So naturally, if you've just bought a call or a put option, you're going to want to see the underlying commodity move. Thus, you'll want a high volatility option. On the other hand, if you're selling an option, you'll want to see the price of the commodity remain fairly stable throughout the life of the option. Of course, you can use this reasoning in reverse. If you think a certain commodity will not be experiencing any considerable volatility in the near future, you might put in place a strategy of selling call or put options. In contrast, if you do think the commodity will go through substantial volatility, you might consider buying call or put options. Of course, even if you're right about the volatility, you might not necessarily be right about the direction of the price move. However, as you'll learn in later chapters, there are strategies you can use - such as *straddle*s and *spreads* - which allow you to profit in both directions! But for now, let's just fully grasp this concept.

One simple way to compute volatility, is to subtract the lowest price from the highest price reached during a twelve month period, and then divide the difference by the annual low, and then multiply by 100:

<u>Highest price in 12 months MINUS Lowest price in 12 months   x 100 </u>
             DIVIDED by Lowest price in 12 months

You may also come across a term called **statistical volatility**. This simply refers to the level of volatility that this commodity (or stock, or index), has shown in the past. Though statistical volatility is taken into account when pricing an option, what really gives the option more value is the likelihood that the price will have big moves in the future, regardless of what it has done in the past. After all, as we all know, the past is not always a predictor of the future. So as you may imagine, an option that's considered highly volatile will cost more that one that has *low volatility*.

This is why you don't always get the whole story just by looking at the price of the option. For instance, the premium you might have to pay for a particular corn option that's *in the money* might appear expensive at first. But if the price of corn is expected to move sharply over the next several weeks, then from a *volatility* point of view the price of this option is actually very cheap. On the other hand, a cheap corn option might appear very appealing, but if the likelihood of the price of corn moving anytime soon is very low, in other words, if it's a low volatility option, then it's not such a great deal as you might have thought.

Delta
This is the relationship between changes in the option's premium and changes in the price of the underlying commodity. In other words, it tells you how much your option will make or lose as the price of the underlying commodity goes up or down.

For instance, say the delta for a March 5.50 soybeans call is 35. This would mean, that as the owner of a call option you have the equivalent of 0.35 of a long futures contract. If the price of soybeans goes up 10 ticks, your option will appreciate 3.5 ticks. If the price of soybeans goes down 10 ticks, then your option would lose 3.5 ticks of value.

Likewise, if the delta of a March 3.30 corn put is 25, as the owner of the put option you would have the equivalent of 0.25 of a short futures contract. So, if the price of corn goes up 10 ticks, your option would lose 2.5 ticks of value. In turn, if the price of corn goes down by 10 ticks, then your put option will appreciate 2.5 ticks.

In our next chapter, we'll begin looking at the trading process in greater detail.

Chapter Eight

Buying a Call Option

If you're of the opinion that a certain commodity will INCREASE in price within a SPECIFIC period of time, then you should consider buying a futures call option.

If the events prove you wrong, and the price of the underlying commodity you're trading the option on either goes down, or doesn't move at all, the maximum you'll lose is the premium you paid for the option, plus commissions and transaction costs. On the other hand, if the events prove you right, you stand to make *unlimited* profits.

BUYING CALL OPTIONS
As the price of the underlying commodity **increases**, the value of a call option **increases**

Before you purchase a call option, you'll need to consider a few things:

First, you need to look at the length of the option. Should you buy an option that's 2 months, 3 months, or 6 months away from expiration? There's no exact answer since every situation is different, and it'll ultimately depend on your strategies and objectives. Just remember, options are a wasting asset. You only have a specific number of months in which to achieve your profit objectives. Otherwise, you lose your premium. The greater the length of the option, the better the chances of the option increasing in value. However, length has an impact on how much you're going to pay for that option. A longer option will normally cost more. And the more you pay for the option the greater your **break-even point.**

This leads us to the next thing you'll want to know: what will the underlying futures price have to be for the option to break even? To calculate this you just need to know 3 elemental things:
1. Strike price of the option.
2. The premium of the option.
3. Commissions and transaction costs.

Once you know these three things, determining the break-even price is easy:

CALL BUYING BREAK-EVEN POINT
Break-even price = Strike price + Premium + Commissions & Transaction Costs

Example:

You buy a wheat call option with a strike price of 350 (*Represents $3.50 a bushel. Wheat trades in bushels per cent, so it's often just abbreviated in the financial newspapers as 350*).

You pay a premium of $400, plus $50 for commission and transaction costs. The current market price of wheat contracts is $3.50 a bushel. Based on these figures, your break-even price is $3.59 a bushel. Here's the math:

1st) Strike price $3.50

2nd) Find out what the $400 premium translates into at cents per bushel. Wheat trades in 5,000 bushels so:
$400 / 5,000 bu = 0.08 = 8 cents

3rd) Find out what the $50 commission and transaction costs translate into at cents per bushel:
$50 / 5,000 bu = 0.01 = 1 cent

4th) Break-even = $3.50 + .08 + .01 = **$3.59**

Next since options trade down to fractional values as small as sixteenths of a point, I thought it would be a good idea to include a fractional values chart even though throughout our examples we'll be rounding off.

Fractional Values

Fraction	Dollar Value	Fraction	Dollar Value
1/16	$ 6.25	9/16	$56.25
1/8	$12.50	5/8	$62.50
3/16	$18.75	11/16	$68.75
1/4	$25.00	3/4	$75.00
5/16	$31.25	13/16	$81.25
3/8	$37.50	7/8	$87.50
7/16	$43.75	15/16	$93.75
1/2	$50.00	1	$100.00

Let's illustrate buying a call with some examples. We'll look at buying call options that are in the money, at the money, and out of the money.

Examples:

Suppose it's February. You're considering buying a June 600 Silver call option because you believe the price of silver is going up (*This means that the option would actually expire sometime in May. Remember, the stated expiration month corresponds to the expiration of the futures contract, not the option. Options expire 1 month earlier*).

COMEX silver trades for 5,000 troy ounce contracts in cents per ounce (*see chapter 14 for price/unit specifications.*) The current market price is 600 cents per troy ounce ($6.00 per ounce). Therefore, this call option is at the money. Right now, it's got no intrinsic value. But since you believe the price of silver will rise sometime between February and May, you pay a premium of $750 (that's 15 cents per troy ounce) plus commissions and transactions of $100 (that's 2 cents per troy ounce), and purchase the call.

Around March, the price of silver is up to 612 per troy ounce. This means that your option is now **in the money**. It has an intrinsic value of 12 cents per troy ounce. You're up $600 (.12 x 5,000 = 600). If you sold it now you would recover your premium, but you would be out of your commission and transaction costs. Your break-even point is actually at 617. Obviously, it wouldn't be profitable for you to liquidate it just yet.

In April, the price of silver increased to 617 per troy ounce. You're up $850. Minus the $750 premium and the commission you paid, you find that you just **broke even**. So now you've got to make a decision. The expiration date is getting closer, and you have a chance right now to liquidate and break even. At this point, if you hold onto it, there're a few possible scenarios:

Scenario a): The price of silver keeps going up and you offset your option **before** the expiration date and **make a profit.**

Scenario b): The price remains pretty much the same and you **break even.**

Scenario c): The price abruptly turns in the other direction, so you offset your option **before** the expiration date and **incur a partial loss.**

Scenario d): The price suddenly turns in the other direction, the option expires and you **lose it all.**

Let's look at how you would handle each scenario:

Scenario a): In early May, the price of silver jumps to 680 cents an ounce. When you had originally purchased this call option it had $0 intrinsic value. Now it has $4,000 of intrinsic value (.80 x 5,000 = 4,000). You quickly liquidate the option and realize a profit of $4,000 minus the $750 premium and the $100 commission and transaction costs you paid.

Scenario b): By the expiration date in late May, the price of silver is holding at 617 cents an ounce. You liquidate the option and make a profit of $850 minus the $750 premium and the $100 commission and transactions costs. In other words, you break even.

Scenario c): In early May, the price of silver begins to turn against you. It went down from 617 to 614, and now it's at 610. You decide enough is enough, it's time to jump off the boat and offset your option before it's too late. When you liquidate your option you make $500 (.10 x 5000 = 500). Minus the $750 premium and commissions paid, you come out losing a few hundred dollars. However, you decide that it was better to take a limited loss, rather than wait for expiration and risk perhaps your entire investment.

Scenario d): Same scenario as in c), only this time your reasoning is different. You feel there might be a late month surge in the prices of silver, and hold onto the option. But instead, prices continue dropping to 580 an ounce. This placed your option **out of the money**. It expires, and you lose $750 plus commissions.

The following chart illustrates what went on in this example. Notice how our call option began gaining intrinsic value once the underlying silver futures contract rose above the strike price of 600:

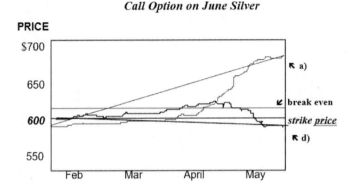

Call Option on June Silver

The a) line is our option climbing to 680 an ounce. The d) line is our option losing ground all the way down to 580 an ounce. In the middle, is our strike price at 600. We figured our break-even point to be at 617.

Defining Your Profit Objectives

You can see by our previous example how important it is to plan ahead, and have definitive profit goals before you start trading an option. As the price fluctuates, you'll be making some decisions. So before you trade, decide:

1. How much of the premium are you willing to lose if things go wrong?

2. How much of a profit are you satisfied with if things go well?

Knowing when to enter and when to exit a trade is key to earning profits!

Let's analyze our example a little deeper.

Suppose it's February and you're very bullish on silver. You're expecting it's going to shoot up over the next few weeks. The price of silver is currently 600 an ounce. You decide on a 350 May silver call option. This means it's already in the money, with an intrinsic value of 250 cents an ounce. That's a $12,500 profit just for buying the call. What a deal!, right? Well, not exactly. Because the premium you're going to be asked to pay for this call will also be higher. In fact, you may pay a 248 cents premium. This translates into $2.48 cents an ounce, which amounts to $12,400 (2.48 x 5,000 = 12,400) plus commissions.

So, even though you have intrinsic value built-into, you're right at about you're break even point depending on your commission fees.

On the other hand, suppose you want to buy a cheaper option. It's February, and you choose a 625 July silver call option. Say the price of silver is currently 525. You're 100 cents out of the money. There's no intrinsic value in this option. However, you feel the price of silver will go up substantially by June (volatility), and you pay a premium of 4 cents an ounce for the option. This amounts to $200 (.04 x 5000 = 200), plus commissions.

Let's say your predictions were right, and the price of silver climbs to 700 by June. Now you're option is in the money, with an intrinsic value of 75 cents an ounce. You liquidate the option and make a profit of $3,750 (.75 x 5000 = 3,750), minus the $200 premium and the commissions. A great return, considering that if had you been wrong about the direction in the price of silver, you would have only lost the $200 premium plus commissions.

Finally, let's close our example with one last scenario. Suppose this time, that instead of buying the 625 July silver call option, you wanted to give yourself more time for the price of silver to rise, so you buy the 675 September silver call option. The market price of silver is still at 525. Your call option, therefore, is 150 cents <u>deep out of the money.</u> Yet, you're asked to pay a premium of 9 cents an ounce. In other words, $450 (.09 x 5000 = 450). The reason you're paying a higher premium with the September option than with the July option, is because you've got a couple of extra months to realize a profit. Simply, the more time you allow, the greater the likelihood that the option will become profitable. With a shorter expiration date, you have a greater risk that the option will expire before the underlying commodity increases in value.

Recapping what we've just covered, we've seen how an option that's in the money commands a higher premium. And the farther away the expiration date is, the higher that premium will be. On the other hand, when an option is at the money, or out of the money, the premium is lower. But just as with the in the money option, the farther away the expiration date is, the higher the premium.

Advantages And Disadvantages of Buying Call Options

The plain fact here is that most call buyers lose money. Even when the underlying commodity increases in value, it may not increase enough. Not only must you be right on which direction the commodity is going to move, but the degree of movement must be big enough to produce a profit within the limited time before expiration of the option. Yet, they do offer some attractive features:

ADVANTAGES

- Limited risk - If the price of the commodity drops, you're total possible loss is the purchase price of the option plus commissions.

- Small capital commitment -The cost of buying an option is substantially lower compared to the cost of a futures contract (or most any other asset.)

- Unlimited profit potential - Even though your risk is pre-defined, your potential for profits are virtually unlimited.

DISADVANTAGES:

- Lower probabilities - It's difficult to consistently guess correctly the timely price direction of the market.

So why do people buy call options? Well, one reason is that it could be part of an overall strategy, as we'll see later on. Another reason, is that even though the odds work against you, it's possible that with a few profitable options you can more than make up for losses incurred with the other not profitable options.

Chapter Nine

Buying a Put Option

If you believe that a certain commodity will be DECREASING in price within a SPECIFIC period of time, then you should consider buying a futures put option.

Just as with call options, if the events prove you wrong, and the price of the underlying commodity moves in the opposite direction, or doesn't move at all, the maximum you'll lose is the premium you paid for the option, plus commissions and transaction costs. On the other hand, if the events prove you right, you stand to make *unlimited* profits.

The primary difference between the put option and the call option, is that with the put option your outlook on the market is *bearish*. While with the call option your outlook is *bullish*. In effect, with put options **we want the price of the commodity to drop.**

BUYING PUT OPTIONS

As the price of the underlying commodity **decreases,** the value of a call option **increases.**

Before buying the put option, you'll need to make the same type of considerations as you made with the call option. That is, what is the length of the option you're trading? And what is your break-even point?

Remember, the greater the length of the option, the better the chances of the option becoming profitable. In the case of a put, the option becomes profitable as the commodity decreases in price. Here again though, length has an impact on how much you're going to pay for that option. A longer option will cost more. And the more you pay for the option the greater your **break-even point.**

A put option is a mirror image of a call option. Everything is just reversed. So when figuring your break-even point for a put option, instead of adding the premium to the strike price, you're going to subtract them.

PUT OPTION BREAK-EVEN POINT
Break-even price = Strike price - Premium - Commissions & Transaction Costs

Example:
You buy a wheat put option with a strike price of 350 (*Represents $3.50 a bushel. Wheat trades in bushels per cent, so it's often just abbreviated in the financial newspapers as 350*). You pay a premium of $400, plus $50 for commission and transaction costs. The current market price of wheat contracts is $3.50 a bushel. Based on these figures, your break-even price is $3.41 a bushel. Here's the math:

1st) Strike price $3.50
2nd) Find out what the $400 premium translates into at cents per bushel. Wheat trades in 5,000 bushels so:
$400 / 5,000 bu = 0.08 = 8 cents
3rd) Find out what the $50 commission and transaction costs translate into at cents per bushel:
$50 / 5,000 bu = 0.01 = 1 cent
4th) Break-even = $3.50 - .08 -.01 = **$3.41**

Just as we've done in the previous chapter with call options, let's look at some examples of buying put options in the money, at the money and out of the money.

Examples:
We're in the month of March, and you're considering buying a 390 Gold June put option. Gold is currently trading at $390 an ounce (*COMEX Gold trades in 100 troy*

ounce contracts in dollars per ounce).

You're bearish on gold, and believe the price of gold is going to drop between March and the actual option expiration date in May (*Your broker should provide you with exact May expiration date*). This put option will give you **the right** (**but not the obligation**) to sell a specified 100-ounce gold futures contract at the strike price of $390 an ounce, anytime before expiration of the option. You pay a premium of $450 for it.

In other words, $4.50 an ounce ($450/100 ounces = 4.50). Plus $100 commission and transaction costs, which translate into $1 an ounce. Your break-even price, therefore, is $384.50. When gold goes down to that price, you can recover all your costs. Anything lower is net profit.

At this time, this put option is <u>at the money</u>. It has no <u>intrinsic value.</u>

Now it's the month of April, and the price of gold has indeed fallen. The current market price is $384.50 an ounce. The option is now <u>in the money</u> and has some intrinsic value built-up. If you would decide to sell the option back and close the transaction, you would break even. So at this junction, here are the possibilities:

Scenario a): The price of gold keeps going down and you offset your option before the expiration date and make a profit.

Scenario b): The price remains pretty much the same and you break even.

Scenario c:) The price suddenly turns in the other direction, so you offset your option **before** the expiration date and incur a partial loss.

Scenario d:) The price suddenly turns in the other direction, the option expires and you lose it all.

Faced with these possibilities, let's follow through:

Scenario a): Before expiration, the price of gold drops to $360 an ounce. You sell the option at the strike price of $390 an ounce in the open market as part of a closing transaction, realizing a profit of $2,450, minus $550 for all your costs.

Scenario b): The price of gold holds steady at $384.50 on ounce up to expiration date. You liquidate the option at the strike price of $390 an ounce and make a profit of $550. When you minus the $450 premium and the $100 commission and transactions costs, you break even.

Scenario c:) Instead of falling, the price of gold rises from $384.50, to $388. You're worried it's getting too close to expiration, and there won't be time left for the option to become profitable, so you decide to limit your losses by selling the option back at $390. You have a $350 loss.

Scenario d:) Same scenario as in c), only this time you hold onto the option all the way to the end, hoping for a reversal of fortune. Instead, the price climbs to 395 an ounce. Your option is now out of the money with no time left. It expires, and you lose $550.

As the price of the underlying commodity - in this case the 100-ounce June gold contract rose, the put option lost all and any of its intrinsic value. Likewise, as it was getting near the expiration date, it was losing its time value.

A graphical representation follows:

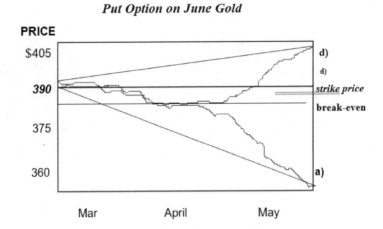

Put Option on June Gold

The **a)** line is our option *gaining* value by *decreasing* to 360. The **d)** line is our option *losing* value by *increasing* to 395. In the middle, is our strike price at 390. We figured our break-even point to be at 384.50.

Let's continue with our example:

It's the month of March, and the price of gold has been sliding for weeks. The market price of a 100-ounce gold contract is currently 370 an ounce. You feel the trend will continue, so you want to buy a short expiration option that's already in the money, and decide on a 390 May gold put option, giving you the right to sell a 100-ounce May gold contract at $390 before expiration sometime in April. It has a built-up intrinsic value of $20 an ounce.

But don't get too excited just yet, because as you know, nothing's for free. The premium on this option will cost you $2,500 or $25.00 an ounce ($2,500 / 100 ounces =

$25.00) Plus $100 for commission and transaction costs (that's $1 an ounce). Therefore, your break-even point is $364 an ounce. The price of gold is going to have to drop at least $6 an ounce over the next month, just for you to break even.

In just a matter of a week, the price of gold falls to $360 an ounce. Since you're a disciplined trader with established profit objectives, you sell the option back and take a $400 profit ($364 - $360 = $4; and $4.00 x 100 ounces = $400). You made this $400 in just one week because you entered and exited the market at the right time. You didn't deviate from your plan and avoided getting too greedy. You also recognized that a small but quick gain, is just as valuable as a larger gain that may be spread out over a longer period of time.

Now let's change the setting just a little. This time you want a cheaper option. We're in the month of March, and you feel the price of gold is going to come down, but you're unsure as to how long it will take. Therefore, you buy an <u>out of the money</u> 360 August Gold put option. The market price of gold is currently 375 an ounce. You're $15 an ounce out of the money. You pay your premium of $160 (that's $1.60 per ounce) and commission and transaction costs of $90 (that's 90 cents an ounce). Based on this, your break-even price is $357.50 an ounce ($360 - $1.60 - .90 = $357.50). Gold will have to drop $17.50 by July, for you just to break even.

Within only two weeks, the price of gold collapses to a year low of $355 an ounce. If you sell back your option now, you'll end up with a net profit of $250 (break-even price of $357.50 - $355 = $2.50, and $2.50 x 100 ounces = $250). That's a 100% return on your investment.

But since you haven't established exact profit objectives, you refuse to exit, and decide instead to ride the option a little while longer. However, after some price oscillations, gold climbs back to $375 and settles. Your option expires and you lose $250 (premium and commissions).

On the one hand, you can view this lost as simply a bad decision on your part. Even though at one point your returns were 100%, your lack of clear profit goals obscured your better judgment as to when to exit the market. You not only lost your costs, but also the opportunity to double your money.

However, since a put option (just as the call option) limits your risk to just the up-front costs of the option, while offering you the opportunity of unlimited gains, you could say it wasn't a bad gamble. After all, you risked a relatively small amount of money.

Advantages And Disadvantages of Buying Put Options

Once again, the plain fact here is that time works against all option buyers (puts and calls). As you get near the expiration date, your put option has a declining time value, and when it expires it becomes worthless.

In our examples, we've selected an in the money high-priced put, very close to expiration, but where we stood to lose a lot more money if the underlying commodity moved in the wrong direction. In contrast, we've seen the purchase of an out of the money low-priced put, farther away form expiration. In this case our capital commitment was much lower, but we needed many points of price movement for a profit.

So it should be clear by now, that the further out of the money a commodity is, the cheaper the premium for a put (or a call), and the lower the chance for profit. And the further in the money a commodity is, the more expensive the put (or the call), because of its intrinsic value. In the same manner, an in the money or an out of the money put (or call) will command a higher premium the farther away it is from expiration.

It should also be quite evident by now, that's its important to enter any trade with clear and concise profit objectives. You should have a pre-defined plan as to when to exit the trade. This will help you immensely. One can argue that if you're going to lose, it's better to lose at least by sticking to the plan than by being side-blinded.

Put option features include these advantages and disadvantages:

ADVANTAGES

- Limited risk - If the price of the commodity rises, you're total possible loss is the purchase price of the put option plus commissions and transaction costs.

- Small capital commitment -The cost of buying a put option is substantially lower compared to the cost of a futures contract (or most any other asset.)

- Unlimited profit potential - Even though your risk is pre-defined, your potential for profits are virtually unlimited.

DISADVANTAGES:

- Lower probabilities - It's difficult to consistently guess correctly the timely price direction of the market.

So why do people buy put options? Just as with call options, while the odds work against you, it's possible that with a few profitable options you can more than make up for losses incurred with the others.

Also, for many people who hold positions in futures contracts (or own shares of stock for that manner), the risk of decline in price is constantly a concern. For protection against such a price decline, buying a put option would act as an insurance policy.

This is known as a **hedge strategy.** Some investors might even purchase one put option for every contract they own. This is known as a **downside protection.** If the value of your futures contract declines, you would exercise the put and sell the futures contract through exercise.

Chapter Ten

Selling a Call Option

So far, we've been through the process of buying a call option when we thought the price of a commodity would increase, and of buying a put option when we thought the price of a commodity would decrease. If our option became profitable, we would then liquidate it before the expiration date, by selling it back in the open market through the same futures exchange, in a process known as an **offsetting transaction**. In other words, when your 'opening' transaction is a buy, your 'closing' transaction will be a sell. And vice versa.

When we bought calls and puts, somebody else was selling them to us. In this chapter, we're going to be on the other side of the trade. We will be the sellers! In other words, our 'opening' transaction is going to be a **sell**, and our 'closing' transaction will be a **buy**. Thus, this kind of selling is different than the one we've been involved with before, where we were selling back an option we had previously bought, in an *offsetting transaction*. Here, as the seller of the option, when we're ready to close out our position, we'll be buying back the option.

The reason you would sell a call option short, is because you're expecting the price of the underlying commodity to either decline or remain unchanged.

> **SELLING CALL OPTIONS**
> As the price of the underlying commodity **decreases,** the value of a call option **decreases.**

In our previous chapter, if we thought the price of a commodity was going to decline, we'd buy a put option. So why not just buy a put option here? Because the benefits and risks of *buying options* are completely different than the benefits and risks of *selling options*. And you decide on which side of the fence you want to be in, based on these benefits and risks.

By **selling short** a futures call, you're giving the buyer the right to buy a specified underlying futures contract from you at the stated strike price. Remember, you don't really own what you're selling. Instead, you're creating a situation where there will be an **offsetting purchase** pending, to offset the position and bring everything back to balance. In other words, if the buyer of the option exercises his right to buy the specified underlying futures contract at the specified strike price, you will have to acquire a futures position at the current market price for that particular futures contract, and then sell it back to the buyer at the strike price of the option. In effect, as an option seller, this would be your offsetting transaction if you lose with the option. The reason selling options carries unlimited risks, is that you have no control over how much the futures contract may increase.

> *When selling a call, the larger the gap between the increased market price and the strike price, the larger your losses.*

On the other hand, if the buyer doesn't exercise his right because the option doesn't become profitable for him, then you don't need to acquire a position in the futures market. You simply let the option expire.

Our break-even point calculation is a bit different than when you're buying calls. Although you add the strike price to the premium, you must subtract your commission and transactions fees. Because here, you make money up to the break-even point (when buying calls, of course, you make money after the break-even point.)

CALL SELLING BREAK-EVEN POINT
Break-even price = Strike price + Premium + Commissions & Transaction Costs

The best way to grasp the concept will be to go over a few examples. We'll assume that we're short selling *uncovered call options*. Therefore, the risk is greater than if we were selling *covered options*.

> ***Note:***
> *Uncovered calls = you don't own the underlying asset.*
> *Covered calls = you own the underlying asset.*

You'll find, as we go along, that **time is on the seller's side**. That's right, the buyer's greatest enemy – disappearing time value – is the seller's greatest ally!

When you sell a call option, you receive up-front money from the buyer in the form of a premium. That is, the price the buyer's paying for the option. This is the same premium we've been discussing in our previous chapters, only this time – as the seller - the premium is paid to you. This money gets credited to your account.

Your profit objective as an option seller is simple: <u>hold onto this premium until the expiration of the option</u>. And you'll manage to do this, as long as the option doesn't become profitable for the buyer and he exercises the option. So as you can see, your gain potential is limited simply to the premium amount the buyer paid you for the option.

Examples:

It's the month of August. Soybeans are trading at $8.50 a bushel. You believe the price of soybeans will either decline or remain the same over the next few months. So you decide to sell a 925 November soybeans call (*the common abbreviation for $9.25 a bushel*). That's .75 out of the money. The November call's premium is 28 (*the common abbreviation of 28 cents a bushel*). This translates to $1,400 (.28 x 5,000 bushels, or $50 per penny). You have commission and transaction fees that come to $100 or .02 per bushel ($100/5,000 = .02). So your break-even price is $9.51 ($9.25 + .28 - .02).

At this point, your brokerage account is credited with $1,400 minus commission and transaction costs:

CREDIT ACCOUNT PREMIUM	$1,400
LESS TRANSACTION COST	- 100
BALANCE	$1,300

After some minor price fluctuations, the price of soybeans settles at $9.00, and the option expires. Since this is below the strike price of $9.25, the option expires worthless for the buyer and he cannot exercise it. Your net profit = $1,300:

CREDIT ACCOUNT PREMIUM	$1,400
LESS TRANSACTION COST	- 100
NET PROFIT	$1,300

Suppose that instead of $9.00, the price of soybeans closes at $9.25 by expiration. Would the buyer exercise his right to buy soybeans from you?

No, he wouldn't. It would be senseless since he's still out his $1,400 plus commissions and transaction fees. Remember, the call buyer's break-even price is higher than the strike price of the option. Therefore, he'll let the option expire worthless. You make $1,300.

To continue, let's say that the price of soybeans closes at $9.30 a bushel. Would the buyer then want to exercise his right to buy soybeans from you at $9.25 a bushel? That's 5 cents a bushel below the current market price. Of course he would. He'll at least recover $250 from the premium and commissions he paid (.05 x 5000 bushels = $250). But you, as the seller, are still the winner. Your brokerage account would look like this:

CREDIT ACCOUNT PREMIUM	$1,400
LESS TRANSACTION COST	100

LESS OPTION INTRINSIC VALUE	- 250 (.05 above strike price)
NET PROFIT	$1,050

But, let's suppose that the price of soybeans doesn't close at $9.30, but rather keeps climbing beyond your break-even point, reaching $9.85. The buyer now exercises his right to buy soybeans from you at the strike price of $9.25. You're obligated to 'make good' on the option, no matter how high it had gotten. This would force you - theoretically -to buy soybeans at the market price of $9.85 and deliver them to the buyer at the strike price of $9.25. As a practical matter though, you would simply offset the option by buying it back and paying the exercise value difference. That is, the 60 cents difference between the market price of $9.85 and the strike price of $9.25.

Your account would look like this:

CREDIT ACCOUNT PREMIUM	$1,400
LESS TRANSACTION COST	- 100
LESS OPTION INTRINSIC VALUE	3,000 (.60 above strike price)
NET LOSS	($1,700)

But here's one **very important point**: as the seller of the option (*also referred to as the "writer" of the option*) you can bail out at any time. In other words, you can cancel your open position any time **before expiration** and **before exercise**. And you would do this, of course, by buying back the same option. You'll still make money if :

a) the price of the underlying commodity has declined.
b) the price of the underlying commodity remained the same.
c) the price of the underlying commodity increased, but is still below your break-even point.
d) the time value of the option declined.

For example:
Continuing with our soybeans, suppose August and September goes by and the price remains at about $9.00. The option will expire in only one more month. This means that time value has shrunk enough that you may now have a choice. <u>You can close out your position by buying back the option at a lower price (the price of the premium at this point is lower), or you can just wait for expiration, in which case you'll keep the entire premium</u>. The pros and cons, are that if you buy back the option now you'll make less money, but you end your risk. You ensure your profits <u>today.</u> On the other hand, waiting

for expiration keeps you exposed to risk until the end of the option, but provides you with a possibility of greater financial reward.

As you can see, selling call options carries a lot of risk. Because of this risk, brokers will require that you maintain a margin in your brokerage account. This will vary from broker to broker, and serves as a protection in case exceptionally high losses are incurred.

One final point, while most calls aren't commonly exercised early, once a call option reaches its strike price it could be exercised by the buyer at any time. Even if the call is still several months from expiration. If this happens, you'd be notified by your broker that your call has been exercised. The way it works, is that the *Options Clearing Corporation* (OCC), acts as a buyer to every seller and as a seller to every buyer. So when a buyer exercises an option, the order is randomly assigned to any seller.

Recapping what we've learned, there are two ways to make money as the seller of a call option:

1. If the price of the underlying commodity remains below the striking price of the option

(or to be more precise, below your break-even point.)

2. If the price of the underlying commodity remains stable and the option declines in time value, allowing you to buy it back at a lower premium.

Advantages And Disadvantages of Buying Put Options

The characteristics of selling a call option are completely opposite of those for buying a call option. Naturally, the advantages and disadvantages are opposing as well. Remember when we've mentioned before, that most option buyers lose money? Well, the opposite is true for most option sellers.

The reality is, that as a seller the odds are in your favor. Big advantage, no doubt!

ADVANTAGES

- Small capital commitment – the cost of selling a call option is substantially lower compared to the cost of buying a futures contract (or most any other asset.)

- Higher probabilities of profit - the odds are automatically in your favor when you sell a call option. The price of the commodity can decline, remain the same, or even go up a little, and you still make money.

DISADVANTAGES:

- Limited profit potential - your potential profits are limited the premium paid to you for the option.

- Unlimited risk - if the price of the commodity increases you're obligated to make good on the option by acquiring an opposite futures position, regardless of how high the price becomes.

So why do people sell call options? Well, selling call options, is spite of its risks, can be highly profitable by providing a steady stream of income through the premiums you collect. The fact is, most call options don't become profitable for the buyer. In fact, it is said that **90% of option buyers are losers**. Either they move in the wrong direction, or they increase in price, but not enough to reach the striking price, and thus expire worthless. Also, as we'll see later, you might sell a call option as part of a strategy, by combining this with some other technique that may allow you to cap your risk.

Chapter Eleven

Selling a Put Option

One more time, we'll examine the process of selling an option, only in this instance we'll be selling a put option instead. Again, our 'opening' transaction is going to be **a sell**, and our 'closing' transaction will be **a buy.** Only this time we're selling options for a different reason.

You sell a put option, when you're expecting the price of the underlying commodity to either increase or remain unchanged.

> **SELLING PUT OPTIONS**
> As the price of the underlying commodity **increases,** the value of a put option **increases.**

Your reasoning, is that if you're correct and the price of the commodity goes up or doesn't move, you'll make money by keeping the premium paid to you for the option. By selling a put option, you're giving the buyer the right to buy a specified underlying futures contract from you at the stated strike price. Remember, you don't really own what you're selling. Instead, you're creating a situation where there will be an offsetting purchase pending, to offset the position and bring everything back to balance.

So - just as with the call option - if the buyer of the put exercises his right to buy the specified underlying futures contract at the specified strike price, you'll have to acquire a futures position at the current market price for that particular futures contract, and then sell it back to the buyer at the strike price of the option. In effect, as an option seller, this would be your offsetting transaction if you lose with the option.

The risks associated with selling puts are similar to those associated with selling calls. Theoretically, one could argue that with a put, the lowest price the underlying commodity could reach is zero. While with a call, the highest price the underlying commodity could reach is unlimited. In practice, this makes little difference since either scenario would mean huge losses to the seller. <u>When selling a put, the larger the gap between the decreased market price and the strike price, the larger your losses.</u>

Our break-even point calculation is exactly reversed

than when selling calls. Because when selling calls we made money <u>up to the break-even point.</u> In contrast, when selling puts, you make money <u>down to the break-even point</u>, since you're working in the other direction. Therefore, you subtract the premium and add the costs of selling.

PUT SELLING BREAK-EVEN POINT
Break-even price =   Strike price - Premium + Commissions & Transaction Costs

Your profit objective as a put option seller is to <u>hold onto the premium you're paid for granting the option, until the expiration of the option.</u> And you'll manage to do this, so long as the price of the commodity doesn't decrease enough to make the option profitable for the buyer. So your gain potential is limited simply to the premium amount the buyer paid you for the option.

Example:
It's s the month of June. Wheat - which trades in 5,000 bushel contracts - is currently trading at $3.50 a bushel. You believe the price of wheat is going to shoot up over the next few months. So you sell a 340 September wheat put, which gives the buyer the right to buy wheat from you   at $3.40 a bushel. But since you expect the price of wheat to go up, you anticipate profiting from the premium you're paid. To grant this right, you're paid a

premium of 22 cents a bushel, equivalent to $1,100 (.22 x 5,000 bushels, or $50 per penny). You have commission and transaction fees that amount to $100 or .02 per bushel ($100/5,000 = .02). So your break-even price is $3.20 ($3.40 - .22 + .02). Which means, that once the price goes below $3.20 a bushel you start losing money.

Your brokerage account would reflect the transaction:

CREDIT ACCOUNT PREMIUM	$1,100
LESS TRANSACTION COST	- 100
BALANCE	$1,000

Suppose that by expiration, sometime in August, the price of wheat rose $3.60. This is above the strike price $3.40, thus the option would expire worthless for the buyer. Your account would reflect your net profit:

CREDIT ACCOUNT PREMIUM	$1,100
LESS TRANSACTION COST	- 100
NET PROFIT	$1,000

Now let's assume, instead, that the price of wheat dropped to $3.45. Though wheat didn't go up as you had suspected it would when you sold the option, it still didn't drop far enough for the option to become profitable to the buyer. In effect, it's still above the strike price of $3.40. In other words, it's still out of the money. Consequently, your net profit is $1,000.

But what happens if the price of wheat falls to $3.30? Now the buyer may exercise his option. In which case, you'd buy the option back at the exercise price of 10 cents. This means, you would end up giving back $500 of the premium (.10 x 5,000 bushels = $500), leaving your account as follows:

CREDIT ACCOUNT PREMIUM	$1,100
LESS TRANSACTION COST	- 100
LESS OPTION INTRINSIC VALUE	- 500 (.10 below strike price)
NET PROFIT	$500

Still not bad. However, if the price of wheat dropped to a much lower level, for instance, $2.50, then you'd suffer huge losses:

CREDIT ACCOUNT PREMIUM	$1,100
LESS TRANSACTION COST	- 100
LESS OPTION INTRINSIC VALUE	- 4,500 (.90 below

strike price)
NET LOSS   ($3,500)

But remember, as the option writer you can buy back the option anytime **before expiration and before it's exercised**, regardless of the current price of the underlying commodity. So you do have a level of control.

To illustrate this further with our example, let's say the price of wheat dropped to $3.43 and you're still a month away from expiration of the option. There's nothing prohibiting you from closing out your position and buying back the option. The time value of the option has shrunk and it's still not in the money, so you'll pay a lower price for the option. The difference between the higher premium you originally received, and the lower premium you just paid to get it back, will be your profit.

In short, there are two ways to make money selling put options:

1. If the price of the underlying commodity remains above the striking price of the option (or to be more precise, above your break-even point.)

2. If the price of the underlying commodity remains stable and the option declines in time value, allowing you to buy it back at a lower premium.

Advantages and Disadvantages of Selling Put Options

When you sell put options, just as in the same matter as with selling call options, the odds are greatly on your side. The problem is, that would you gain in the form of probabilities, you give up in the form of risk. As you've been seeing through our examples, selling options does expose you to unlimited risks.

ADVANTAGES

- Small capital commitment -the cost of selling a put option is substantially lower compared to the cost of buying a futures contract (or most any other asset.)

- Higher probabilities of profit - the odds are automatically in your favor when you sell a put option. The price of the commodity can decline, remain the same, or even go up a little, and you still make money.

DISADVANTAGES

- Limited profit potential - your potential profits are limited the premium paid to you for the option.

- Unlimited risk - If the price of the commodity decreases you're obligated to make good on the option by acquiring an opposite futures position, regardless of how low the price becomes.

So why do people sell put options? Again, in spite of the risks, selling put options can provide you with a steady flow of income with the premiums you collect on every option. Most put options don't become profitable for the buyer. This explains why 90% of option buyers are losers, whether they're buying calls or puts. Either the commodity moves in the wrong direction, or it decreases in price, but not enough to reach the striking price within the life of the option, and thus expire worthless. You might also sell a put option as part of a strategy, by combining this with some other technique that may allow you to cap your risk.

On my next chapter, we'll review these strategies.

Chapter Twelve

Options Trading Strategies

So far, we've seen the four primary ways of trading options: buying calls, selling calls, buying puts, and selling puts. We've uncovered the risks and possibilities associated with each of these trading forms.

Now we're going to see how interesting it gets, as you begin combining these basic trading themes in pursuit of a specific strategy. In this chapter, we're going to explore some combined option techniques.

As an investor (no matter what type of asset you're investing in), it's critical that you understand not only what your profit potentials are, but also the risks involved in any investment. In trying to achieve your profit goals, you should be able to identify what acceptable levels of risks you're willing to assume. This, of course, will depend on your personality, point of view, tolerance for risk, perception of the market, financial position, and overall financial plan.

People trade the markets using different option strategies for entirely different reasons. In essence, they're trying to satisfy different goals.

For instance, as a conservative type trader, you might be looking to make profits only at the cost of assuming minimum risk. You then pursue an option strategy that reflects this philosophy, by putting a cap on your gain potential while also substantially limiting your exposure. Or another example, to protect the value of a long position you've acquired in the futures market, you might buy a put. This will give you downside protection (*a form of 'insurance' in case of a drop in the price of the commodity*) without the need to risk exercise. On the other hand, as a more speculative trader, you might be much more attracted to cheaper, near-expiration options that offer greater profit potential with greater risk.

Whatever your position may be, there's probably a suitable option strategy for you. So let's begin by going over some basic definitions, and then follow up with our examples.

There are 3 major types of combined strategies:

1. **Spreads** – Buying an option while selling another option of the same underlying commodity, with different strike prices and/or different expiration dates. This combination of buying and selling provides a way of putting the odds in your favor, while reducing risks in case the price of the underlying commodity moves adversely. There are different varieties of spreads, and they can get very complex. I'll go over the most common and simple ones here in this chapter.

2. **Straddles** – Simultaneously buying and selling a call and a put for the same underlying commodity, with identical strike prices and expiration dates. The key here is that both the striking price and the expiration date of the call, or of the put, is the same.

3. **Hedge –** To protect one position with another position. For instance, buying a put option to protect your long futures position in case of a price decline. Basically, the premise is that every trade that you make is offset with another trade that has the potential to earn you money if the market turns against you. Again, this is a form of insurance. You're protecting your long position by covering yourself with a strategy that caps your risk in case of a price decline. Conversely, you protect your short position by covering your-

self with a strategy that caps your risk in case of a price increase.

Let's now take a closer look at each one of these techniques, by putting them through examples that can help you grasp the concept.

SPREADS

As I've mentioned earlier, spreads can be used in a variety of ways, and some of these forms of spreads are advanced strategies that involve a more complex analysis. But don't think that just because an option strategy is more advanced and difficult, it has better odds than a more simple strategy. On the contrary, a simple strategy could be just as effective. Besides, more advanced strategies may involve more risk, thus better left to the more experienced trader.

Before we go on with examples, let's explain the two basic categories of spreads: bull spread and bear spread.

Bull Spread: buying and selling a call option, or buying and selling a put option, where your profits increase as the value of the underlying commodity rises. In a bull strategy, you **buy** an option with a **lower** striking price, and **sell** an option with a **higher** striking price.

Bear Spread: buying and selling a call option, or buying and selling a put option, where your profits increase as the value of the underlying commodity drops. In a bear strategy, you **buy** an option with a **higher** striking price, and **sell** an option with a **lower** striking price.

So there are really four possible forms of spread:

Bull spread using calls.
Bull spread using puts.
Bear spread using calls.
Bear spread using puts.

Examples:

(*For simplicity, we won't calculate commissions and transaction costs in our examples*).

Bull Spread - You're following corn prices through the newspapers. It's now the month of June, and corn is currently trading at $2.75 a bushel. There's a seasonal tendency for corn prices to rise into July and August, and it generally bottoms around November. Expecting that this tendency will continue, you decide to implement a bull spread strategy. So you buy an August $3.00 call option on corn, paying a premium of 4.5 cents, which translates into $225 (.045 x 5,000 bushels = $225).

What you're really doing here is buying one out of the money call, and selling two further out of the money calls. At this point, this trade isn't costing us a penny (once again, commissions and transaction costs calculations are left out for simplicity.)

So let's look at the possibilities with different scenarios:

Scenario 1. The price of corn doesn't move (or declines), so therefore we haven't made nor lost anything. In fact, this situation will hold true all the way up to the break-even price of $3.04 1/2. **Outcome = We broke even** (however, we'd be out on our brokerage fees.)

Scenario 2. The price of corn goes up to $3.20 by the end of June. We decide to exit our position early with offsetting transactions. That is, we buy back the options we sold and sell back the option we bought. By selling back the option we bought we'll make $1,000 (.20 x 5,000 bushels = $1,000), minus our cost of $225, equals a net profit of $750. To this amount, we should add the money we collected on the premiums for the options we sold. However, since we're buying back the options we sold a little early, we're giving up part of the premiums. Suppose that to buy back the options we pay a premium of 1.25 cents each (remember, at this point the option has lost time value.) That's a total of $125 (.0125 x 5,000 bushels = $62.50, multiplied by 2 options = $125). Since we had received $225 when we sold the options, we're still ahead by $100.

Outcome = Total net profit from this bull spread strategy = $850.

Scenario 3. The price of corn goes up to $3.30 by expiration time. Therefore, before expiration we sell back the call option we bought for a profit of $1,500 (.30 x 5,000 bushels = $1,500), minus our premium of $225, equals a net profit of $1,275. PLUS, the two call options we sold expire worthless for the buyer, so we keep the $225 premium money we had collected.

Outcome = Total net profit = $1,500.

Scenario 4. The price of corn reaches $3.60 by expiration date. Now, on the option you bought, you'll make $3,000 (.60 x 5,000 bushels = $3,000) minus the $225 premium you paid = $2,775 net profit. On the options you sold, you'll lose $3,000 (.30 x 5,000 bushels = $1,500, multiplied by two options = $3,000), minus the $225

premium you collected = $2,775 loss.
Outcome = We broke even.

As you can see, after the $3.30 level it was dollar for dollar. That is, we lost a dollar for every dollar we made. As the price of the commodity rises we make money with the call option we bought, up to the point when we start losing money with the call option (or options) we sold. Thus, our profits are limited (in this case, up to $1,500); but so are our losses!

One side note, we've could have sold just one call option instead of two. Thus, we would have only collected one premium for $112.50. Our total exposure for the entire transaction would have been limited also to $112.50. By altering our spread a bit and selling two options instead of one, we altered the possible outcomes under each one of our scenarios:

	SELLING TWO OPTIONS	SELLING ONE OPTION
1. ☐	Broke even	Net Loss = $112.50
2. ☐	Net Gain = $850	Net Gain = $800
3. ☐	Net Gain = $1,500	Net Gain = $1,387.50
4. ☐	Broke even	Net Gain = $1,387.50

Of course, we could have entered this bull spread using puts instead. Our thinking would have been that the price of corn was going to be declining over the next few months.

So back to our corn example, the price in June when we entered the trade was $2.75 a bushel. If we wanted to use puts, we would buy an August $2.50 put option, and sell two August $2.20 put options. That is, we're buying an out of the money put, and selling two further out of the money puts. The math is exactly the same, only we're going in the opposite direction.

Naturally, there are variations on any bull spread. As we mentioned earlier, we could have just bought one *out of the money call* and sold one f*urther out of the money call.* Or, we could sell an *in the money put* and buy an *out of the money put.* In this case, we'd collect a higher premium for the put option we sell, because it's in the money already. If the price of the commodity rises, your short position will lose its premium value at a faster rate than your long put, so you would then make profits from your long position.

Bear Spread – Let's say you're looking at soybeans. It's the month of September, and soybeans are trading at $7.00 a bushel. The seasonal tendency is for beans to decline in price in the autumn and rise in price in the spring. You believe this tendency will hold true this year, especially since you've read that this year the carryover percentage is 40%, which is quite high. So you decide to implement a bear spread strategy. You sell a November $8.00 call option for soybeans, and receive a premium of 15 cents, that is, $750 (.15 x 5,000 bushels = $750.) At this point, $750 is considered unrealized gain, and your risk is technically unlimited.

You're expecting the price of beans to fall. Since this call option will actually expire at the end of October, your total time risk is only about two months. But just in case you're wrong about the price of beans falling, you want to protect your downside. So you apply a bear spread and buy a $8.25 November call option for 7 cents. That is, $350 (.07 x 5,000 bushels = $350.) At this point, you're making $400 in profits.

Let's review the possibilities:

Scenario 1. The price of soybeans declines, remains the same, or increases, but not beyond the strike price of $8.00. Thus, you make $750 from the call option you sold, and lose $350 from the call option you bought. **Outcome: Your net profit = $400.**

Scenario 2. The price of soybeans goes up to $8.25. You

lose $1,250 (.25 x 5,000 bushels = $1,250) minus the $750 premium = $500, plus the $350 from the option you bought = $850. This is your maximum possible net loss, and hence, your worst case scenario. Remember, as the price of soybeans rises, you could close your position any time before expiration and avoid letting your short position gain value. But if the price jumps too quickly, you might not have time to do this. In this case, it reached $8.25 and the buyer exercised his option.

Scenario 3. The price of soybeans goes beyond $8.25, up to $9.00. Now the money you're making from the call option you bought is $3,750 (.75 x 5,000 bushels = $3,750) minus the premium of $350 = $3,400. While from the option you sold, you'll lose $5,000 ($1.00 x 5,000 bushels = $5,000), minus your $750 premium = $4,250.

Outcome: Your net loss is $850.

As you see by this example, if you had just outright sold the soybeans call option your maximum potential gain would have been $750, while your maximum potential losses could have been unlimited. By applying a bear spread, you limited your maximum potential gain to $400, but also limited your maximum potential losses to $850.

Why would you do this? Because, what market analysts have found to be true over the years, is that the <u>odds of the underlying commodity reaching the strike price before expiration is less than 20%.</u> Therefore, there's over an 80% chance that scenario 1) from our example would repeat itself. So in our example, it comes down to this: <u>would you risk up to $850 at a 20% chance of losing, in order to earn $400 at an 80% chance of winning?</u> There's really no right answer. This is a personal decision. In a way, it's like being in the insurance business. You collect money from the insurance policies, but there's a 1 in 5 chance of having to pay a large claim.

And of course, we could have entered this bear spread strategy by using puts. Again, there's no difference in the math, we're just betting in the other direction.

Vertical Spreads
When you're trading options that have <u>different striking prices, but identical expiration dates</u>, you're involved in a vertical spread (*also known as a "money spread"*).

This is what we've just been doing in our previous examples. In other words, a vertical spread can be a **vertical bull spread** or a **vertical bear spread**. You can use vertical spreads to take advantage of price changes in the premium value. Generally speaking, when an option is in the money, its price will probably change much faster than when it's out of the money. On this assumption, you open offsetting long and short positions just as we've done with our previous corn and soybeans examples. You expect your position that is in the money (whether it's the long or short one) to increase or decrease in price faster than the offsetting option.

Credit Spreads

A spread position can be entered not only to limit risks, but also to generate profits. A credit spread is a spread that generates a positive cash flow. That is, one in which more cash is received through premiums collected on short positions than paid out on premiums for long positions (*the reverse is called a 'debit spread'. That is, when more cash is going out than coming in through premiums*).

We saw this in our beans example. We received $750 on our short position, and paid out $350 for our long position, a classic credit spread. On the other hand, with the corn, we paid out $225 for our long position, while only receiving $112.50 per premium on our short positions, which would qualify it as a debit spread.

In a credit spread, you're basically using a portion of the premium you've collected to cap your risk. Going back to our insurance company analogy, insurance companies use a service called reinsurance to control risk. That is, they take a portion of the premium money they collect from their policies, and then turn around and purchase insurance of their own. This puts a dent in their profits, but it also limits their exposure in case of a large claim. Their losses won't be devastating.

STRADDLES

In spreads, we buy and sell options with different terms (t*he terms of the option are the striking price, expiration date and underlying commodity)*. But with straddles, we simultaneously buy and sell options with the same terms. There are two types of straddles: long straddle and short straddle.

Long Straddles

You simultaneously buy a September $3.00 call option and pay a premium of $400, and a September $3.00 put option and pay a premium of $250. Your total investment comes to $650. Assuming the underlying commodity your trading is a grain market that trades in 5,000 bushels, your $650 investment would come to 13 cents a bushel ($650 /5,000 bushels = 013.) Therefore, for this straddle to be profitable it must move beyond 13 cents in either direction.

Short Straddles

You simultaneously sell a September $3.00 call option and receive a premium of $400, and a September $3.00 put option and receive a premium of $250. Your unrealized gain comes to $650. Once again, the underlying commodity you're trading is a grain market that trades in 5,000 bushels, so your $650 gain represents a gain of 13 cents a bushel ($650 /5,000 bushels = 013.) Therefore, for this straddle to be profitable it must remain within 13 cents in either direction from the strike price.

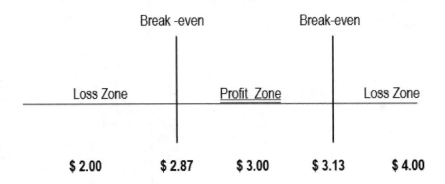

One flaw in this strategy is that, realistically, during the life of the option it's likely that one of the positions in your straddle will be in the money, and therefore worthwhile to exercise. If this happens, a thin margin of profit could be absorbed in commissions and transaction fees.

So, how about if we **COMBINE** our straddle with a spread?

This could be done in different ways. Say, for instance, we want to buy an *at the money* call in corn and sell an *out of the money* call in corn. But we want some protection in the event the market turns against us, so we additionally buy an *at the money* corn put and sell an *out of the money* corn put. This not only reduces our exposure, but increases our chances of making some profits. Though the profits may be less.

Example:

December corn is currently trading at $3.00. You go ahead and buy a $3.00 corn call option which costs you 10 cents a bushel. In other words, your exposure right now is $500. To protect this investment you sell a $3.10 corn call option for 7 cents a bushel, which means you collected a premium of $350. At this point, you reduced your exposure to $150. (So far, this is a debit spread. If you recall, a debit spread is when the amount of premium you paid out exceeds the amount of premium money you received.)

But what if the price of corn drops? Well, to protect this side of the market you buy a $3.00 corn put option. Suppose it's also going for 10 cents. That's another $500 premium you pay. And yes, to protect this exposure on the down side you sell a $2.90 corn put option, which we'll say is also trading at 7 cents. So you receive another premium of $350, which helps recaptures some of your investment.

Your total pay out has been $1,000, and your total collections have been $700. Therefore, your total risk is $300. Here are the possibilities:

Scenario 1. If corn moves up to 6 cents in either direction, you'll break even (.06 x 5,000 bushels = $300).

Scenario 2. If corn moves from 6 cents up to 10 cents in either direction, you'll profit. Your maximum potential profit = $500 (.10 x 5,000 bushels = $500.)

Scenario 3. If corn moves beyond the 10 cents range in either direction you'll break even. Any profit that would have been realized on the $3.00 call is given up to the buyer of the $3.10 call. Likewise, any profit that would have been realized on the $3.00 put is given up to the buyer of the $2.90 put.

Scenario 4. If corn moves less than 6 cents in either direction, you'll incur partial losses. OR, if corn doesn't move at all, you'll lose it all. Total loss = $300.

We used a straddle by buying a $3.00 call option and selling a $3.00 put option. This was intermingled with our debit spread. Though we couldn't guarantee not losing any money, with the straddle we were able to reduce our risk from the original $500 to just $300. Keep in mind, though, the price of the commodity would have to move enough to cover our brokerage fees. The key is to hedge the trade so that we spend the least amount of money possible. This could usually be done by selling further out of the money options, which is a technique called strangle. The real secret here, is to first look at the trade and see how much the total trade will cost you, and then figure out how much you can earn.

In our example, the trade cost us $300. Our maximum potential profit was $500. This is a 3:5 risk/reward ratio, which for a lot of people might not be too appealing. But suppose, instead, that we can trade with a little more leverage and can capture a larger potential profit. Let's say we trade soybean oil, where a penny move is $6.00. In soybean oil the options are priced at 1/2-cent moves.

Therefore, purchasing a 22.50 soybean oil call for 20 cents and selling the 23.00 soybean oil call for 10 cents, allows for a potential $300 profit (50 cents x $6.00 = $300). How much did it cost us to get into the trade? Only $60 (20 cents x $6.00 minus 10 cents x $6.00 = $120 - $60 = $60.) This is a 5:1 risk/reward ratio, a bit more worthwhile than our previous 3:5 ratio.

Of course, this trade can be done in any commodity market. And the closer to expiration, the better. Again, the important thing when zeroing in on a commodity market is to look at all the possibilities. Sometimes the trade won't work no matter what you do. Other times, it's like looking a gift horse in the mouth. Simply pick up the paper or look it up online and do the math on the front month options (*see how to read options in chapter 14*). Your broker can tell you exact expiration dates.

Short Strangle / Long Strangle

When you simultaneously buy an out of the money call and an out of the money put on <u>the same commodity</u> and with <u>the same expiration date,</u> you are said to be **long a strangle**. On the other hand, if you simultaneously sell an out of the money call and an out of the money put on the <u>same commodity</u> and <u>with the same expiration date</u>, you are said to be **short a strangle**. In both cases, the strike price is not the same.

Example 1

It's July 1st, and oats are trading at $2.00 a bushel. You simultaneously sell an August 2.15 call and receive a premium of $300, and sell an August 1.88 put and receive a premium of $400. This is a short strangle with a total profit potential of $700. The probability that the market price of oats will stay within your profit range in this short period of time is about 80%.

However, if the price moves in either direction beyond your profit zone, you could technically incur unlimited losses:

Example 2

It's July 1st and oats are trading at $2.00 a bushel. You simultaneously buy an August 2.15 call and pay a premium of $300, and buy an August 1.88 put and pay a premium of $400. This is a long strangle with a total risk commitment of $700. The probability that the market price of oats may break away from your loss zone is just about 20%. However, if it does, your gain potential is unlimited:

Now, let's COMBINE a strangle with a credit spread.

Example

On May 1st the price of gold is $365 an ounce. We sell a $380 June gold call option for $1,000 ($10 an ounce) and simultaneously buy a $390 June gold call - which is one strike price further out of the money - paying $800 for it. At the same time, we sell a $350 June gold put for $1,200 ($12 an ounce), and simultaneously buy a $340 June gold put - which is one strike price further out of the money - for $900. At this point, our credit = $2,200 for premiums collected, minus $1,700 for premiums paid. Total net credit = $500.

This transaction is a credit spread comprised of two parts: the buying part, which qualifies as a long strangle, and the selling part which is the short strangle. Let's examine some of the possibilities:

Scenario 1. The price of gold remains unchanged, OR moves up to $15 in either direction. **Outcome = we make maximum profit of $500.**

Scenario 2. The price of gold moves beyond $15 and but just below $20 in either direction. **Outcome = we make partial profits.**

Scenario 3. The price of gold moves $20 in either direction. **Outcome = we break even.**

Scenario 4. The price of gold moves beyond $20 but just below $25 in either direction. **Outcome = we incur partial losses.**

Scenario 5. The price of gold moves between $25 and infinity (or $25 and $0). **Outcome = we incur maximum losses of $500.**

In this example, our maximum potential gain was $500, while we limited our total potential risk to only $500 as well. However, the key here was not only having limited our potential losses, put having put the odds at our favor. Historically, bearing an unanticipated catastrophe, the price of this commodity wouldn't swing such a large range, in such a short period of time. Thus, putting the probabilities of profits heavily on our side.

Now let's look at a comparative chart, showing what might have happened in our previous gold example, if we had entered this trade on just a long strangle, or on just a short strangle:

	Spread/Strangle Combined	Long Strangle	Short Strangle
1. ☐	Maximum gain $500	Lose $1,700	Maximum gain $2,200
2. ☐	Make partial gains	Lose $1,700	Make partial gains
3. ☐	Break even	Lose $1,700	Make partial gains
4. ☐	Limited losses	Lose $1,700	Limited gains
5. ☐	Maximum loss $500	Unlimited gains	Unlimited losses

By this chart, it's easy to see that as long as the market behaves the way we historically would expect, the short strangle - in spite of its unlimited risk potential - is an attractive strategy. On the other hand, if you are a risk-adverse type of person, using just only a short strangle wouldn't be your cup of tea, since if the improbable does occur your losses could be theoretically unlimited. Therefore, by using a credit spread, you'll take a bite off of your profit potential, but put a lid on your losses.

HEDGING

Hedging can protect a long or a short position in the underlying commodity, or it can serve to reduce risks in other option positions. Whenever options are bought or sold as part of a strategy to protect another open position, it's described as hedging. If you're in <u>a long hedge you're protecting against price increases.</u> While in a <u>short hedge, you're protecting against declining prices.</u>

By using some of the various forms of combining spreads, straddles and strangles, as we just did in our previous examples, we were engaged in a form of hedging.

Variable Hedging

In a variable hedge, you'll have an uneven number of option positions open. For instance, you might buy two calls and sell one call at a lower striking price. By doing this, you will:

a) Get back some or all of the premium money you put out.

b) Have a chance of profiting from the short call.

c) Offset any potential loss from the short call with potential gains from the long calls.

Example:

Suppose the price of corn is currently $3.40 a bushel. You buy three September $3.50 corn calls at $200 each, for a total of $600. You also sell one September $3.40 corn call (at the money) and receive $750. Thus, your account is credited with net proceeds of $150 ($750 - $600 = $150.) If the price of corn rises above the striking price of $3.50, you'll begin profiting from the three calls you bought at a ratio of 3:1. That is, you'll make $3 from your long position for every $1 you lose from your short position. The more the price of corn increases, the more you'll profit with this variable hedge position. However, you can still profit with your short position. Let's examine some of the possibilities:

Scenario 1. The price of corn drops to $0 or remains unchanged. Therefore, you keep the $150 premium money which was credited to your account earlier.
Outcome = You make $150 profit.

Scenario 2. The price of corn goes up to $3.50. This of course triggers the buyer of your $3.40 call option to sell back the option, making 10 cents on the option. On the other hand, the three call options you've purchased are still worthless since they're just now at the money.
Outcome = You incur maximum losses of $350 (.10x 5,000 bu = $500; $500 - $150 = $350.)

Scenario 3. The price of corn goes up to $3.53 1/2. You lose $675 from your short position (.1350 x 5,000 bu = $675). However, you gain $675 from your long position (.0350 x 5,000 bu = $175; $175 x 3 options = $525; $525 plus the $150 premium = $675.)
Outcome = You break even.

Scenario 4. The price of corn goes up to $3.70. You lose $1,500 from your short position (.30 x 5,000 bu = $1,500.) However, you make $3,150 from your long position (.20 x 5,000 bu = $1,000; $1,000 x 3 options = $3,000; $3,000 plus the $150 premium = $3,150.)
Outcome = You make $1,650 profits ($3,150 - $1,500 = $1,650.)

Scenario 5. The price of corn jumps to $4.00 a bushel. You lose $3,000 from your short position (.60 x 5,000 bu = $3,000.) However, you earn $7,650 from your long position (.50 x 5,000 bu = $2,500; $2,500 x 3 options = $7,500; $7,500 plus $150 premium = $7,650.)
Outcome = You make $4,650 profits. ($7,650 - $3,000 = $4,650.)

As you can see, as the price of the commodity rises you'll make more and more money. Also, notice your narrow losing zone. In this case, anywhere from $3.41 to $3.50, where you can lose a maximum of $350.

In other words, your total risk with this hedge strategy was $350. Your losing zone was narrowed to a 10 cents range. And your profit potential was unlimited.

Of course, you could have gone the other way and have taken on a much more aggressive hedging position, by assuming a short variable hedging position.

Example:
With the price of corn currently at $3.40, you sell four September $3.50 corn calls for $500 each. You also buy two September $3.60 corn calls for $150 each. This leaves a $1,700 credit in your account ($2,000 - $300 = $1,700.) Obviously, your profit potential on the short side will be greater with a short variable hedge than with a long one. In this case, your total potential profit is $1,700. But if the price of the commodity increases, your losses will increase the higher the price rises. Here's the scenario:

Scenario 1. The price of corn declines, remains the same, or increases up to $3.50.
Outcome = You make a $1,700 profit.

Scenario 2. The price of corn goes up to $3.60. Therefore, your four call option buyers exercise their option and you lose $2,000 on the short side.
Outcome = You lose $300 ($2,000 - $1,700 = $300.)

Scenario 3. The price of corn goes up to $3.70. Therefore, you lose $4,000 from your short position. However, you offset part of these losses with your two long posi-

tions (.10 x 5,000 bu = $500; $500 x 2 options = $1,000; $1,000 plus the $1,700 premium = $2,700.)
Outcome = You lose $1,300 ($4,000 - $2,700 = $1,300.)

Scenario 4. The price of corn jumps to $4.00 a bushel. You're losing 50 cents per option on your four short positions, and earning 40 cents per option on your two long positions.
Outcome = You lose $4,300 ($10,000 - $5,700 = $4,300.)

Clearly, as the price of the commodity rises, your losses start adding up to the tune of a $4 loss for every $2 gain.

***

OTHER COMBINATIONS

Box Spreads

We've seen the bull spread and we've seen the bear spread both working separately. How about when you use them both working together? In that case, you'll have what is known as a box spread. In other words, when you open a bull spread and a bear spread at the same time and on the same underlying commodity, you're said to be opening a box spread. With a box spread, you'll achieve maximum profits whether the price of the underlying commodity increases or decreases in value.

Example:
You're looking to open a box spread on CBOT wheat. The price of wheat is currently $3.55 a bushel. First: you decide to open a bull spread by using puts. So you sell a December 3.55 put and receive a premium of $600, and you buy a December 3.50 put and pay a premium of $250. Your account is credited with + $350. Second: to complete the box spread you open a bear spread position, and decide to do it with calls. So you buy a 3.60 call for $450, and sell a December 3.70 call for $300. Your debit balance on the bear spread is $150.

Here are the scenarios:

Scenario 1. If the price of wheat rises between 3.55 and 3.60, you close the bull spread at a profit of $350.

Scenario 2. If the price of wheat rises between 3.60 and 3.70, you can close your bear spread at a profit as soon as it passes your break-even point.

Scenario 3. If the price of wheat rises above 3.70, the call you bought and the call you sold will rise dollar for dollar.

Scenario 4. If the price of wheat declines from 3.55 to 3.50, you'll earn limited profits from your bull spread position.

Scenario 5. Once the price of wheat goes below 3.50, the put you sold and the put you bought will move dollar for dollar.

As you can see, as the price of the commodity moves in either direction, you can close either the bull or the bear part of the box spread for a profit. If the price of the commodity suddenly reverses, you'll profit from your other positions. With a box spread, your total profit potential and total risks are based on the balance of your total credits and total debits from the option premiums.

Calendar Spreads

In our examples, we've been using vertical spreads by entering options with different striking prices but with identical expiration dates. If we entered options on a spread with different expiration dates, then we would be using a calendar spread (also known as "Time Spreads"). There are two types of calendar spreads: Diagonal and Horizontal.

Horizontal Spread

You use options that have the <u>same strike prices</u> but <u>different expiration dates.</u>

Example:
Suppose it's January and wheat is going for $3.30 a bushel. You decide to enter a horizontal spread using puts. First, you sell a March 3.00 corn put for $300. Second, you buy a May 3.00 corn put for $500. Thus, this entire operation cost you $200. If your March option expires worthless, you will have made $300 from your short position. And even though you're still down $200 from your long position, you still have 2 months for this option to become profitable.

On the other hand, if the price of wheat falls below the 3.00 level, your long position protects you dollar for dollar from the risk on the short position. Your total risk, therefore, would be the $200. And should the price continue falling beyond March, you'll begin to enter the profit zone from your May long position.

Finally, if the price of wheat rises, you'll make $300 from your March short position, but you'll lose $500 from your May long position. The purpose of this horizontal spread, is that if you had just entered this trade by buying a May 3.00 put and nothing else, your total risk would have been $500. The attraction of the horizontal spread, is that it reduced your exposure from $500 to only $200.

Diagonal Spread

You use options with different strike prices and different expiration dates.

Example:

Using the same wheat example as before, you decide to enter the trade creating a diagonal spread instead. Let's say you want to trade call options. You sell a 3.40 March wheat call and collect a premium of $400. Then you buy a May 3.50 call and pay a premium of $200. At this point, your profit is $200.

If the price of wheat falls, you'll make a $400 profit from your March short position, and lose $200 from your May long position. Therefore, your net profit is $200.

If the price of wheat rises, you'll begin to lose from your March short position, but your May long position will increase in value. Your maximum risk is $300 (.10 x 5,000 bu = $500; $200 - $500 = -$300.)

Finally, after the March short call expires, you still have two months left for your May call to become profitable.

In conclusion, entering this diagonal spread allowed us to narrow our loss zone to only a 10 cent range, while limiting our possible losses to only $300. In the same manner, it allowed us to profit in both directions: On the short side, up to $200 net profit. And on the long side, profits could be potentially unlimited.

OTHER THEORIES

Elliot Wave Theory
The Elliot Wave theory is based on the premise that all natural phenomena are cyclical, and market behavior trends and patterns follow this cycle. Prices unfold in "five waves" of crowd psychology when moving in the direction of a primary trend. Then they move against the trend in "three waves." The wave patterns reflect life's starts, false starts, stops and reversals. By iden-

tifying the exact position of the current price activity within the wave patterns the trader should anticipate the market's next move. Once you're synchronized with the wave patterns, you can ride the economic waves of the markets. While the Elliot Wave has a group of devoted followers, it doesn't have much credence with professional statisticians.

Gann Numbers

In his 1942 book, How to Make Profits Trading Commodities, W.D. Gann advanced his ideas of precise mathematical and geometric predictive patterns that govern, among other things, the commodity markets. He believed these patterns could be uncovered and exploited in order to profit from the markets. An important basis for his trading system are the Fibonacci Numbers.

Fibonacci Numbers

Although reputedly designed by Fibonacci, a 13th century Italian mathematician, as a study exercise for his students, the Fibonacci numerical progression has received a great deal of attention from many technical analysts. The numerical series is created by starting with 1, with each succeeding number being the sum of the preceding two numbers: 1, 2, 3, 5, 8, 13, 21, 34, 55, 89, etc. Using these numbers in various calculations as divisors or multipliers generates a ratio of 1.618 or 0.618.

This was called the golden ratio by the ancients because it appears widely in nature (branching of trees, flowers, sea shells, etc.) and is appealing to the eye. It was used in structures such as the Egyptian pyramids and the Sistine Chapel. As a result, technical analysts reason that if the golden ratio exists in nature, is used by mankind, and is pleasing in formation, it should appear in price charts as well. When it does, they give special attention to price advances or retracements that reflect the golden ratio.

I'm not quite sure what Fibonacci would have thought of all this, since his number progression was originally constructed from observations on the incestuous copulation of rabbits!

Throughout this chapter, we've seen a number of different possible strategies for trading options. To be clear, you can construct them in many different ways. Each technique or combination of techniques serves a specific purpose. Either to lower your investment risk, limit your possibilities of losses, increase your odds of winning, and/or increase your profit potential. In any case, it's tremendously rewarding to devise and execute a well-planned strategy. And as you've seen, you don't even need to know whether the price of a commodity is going to go up or down, since with some of these techniques you can profit in both directions.

What you do need to know for sure, is exactly how much you're risking and exactly how much you can make. And just as importantly, what are the odds of winning or losing. The name of the game here is <u>knowledge</u>. And believe it or not, you've just gained more knowledge than most of the people who are actually currently trading futures or stock options!

But the learning process is one that never ends. As you can imagine, it's an ongoing process and you should continue to educate yourself every opportunity you can. Whether your interest lies in trading commodities by using one of the many technical analysis tools, or by trading by fundamentals, or by trading options, you need to continue adding to your knowledge base. Remember, your mind is like a parachute, it only works when it's open. Every new piece of information you take in, is like adding another building block that will reinforce your structure of knowledge. Ultimately, knowledge is what differentiates the winning trader from a losing one.

The next step is to learn to read option charts. Chapter 13 is dedicated to showing you the short cut, and easiest ways to read and interpret options pricing for all futures markets.

Chapter Thirteen

How to Read Options Pricing Charts

It seems that one of the most confusing things for beginning traders (and for more experienced ones as well), is accurately interpreting the options pricing data commonly obtained from specialized financial sites and papers, such as *Investor's Business Daily* and *Wall Street Journal.* You can imagine how difficult it would be to make timely decisions, if you aren't sure of how to read the quotation tables.

This section is going to help you become efficient at interpreting pricing data. You'll learn the short-cut way to calculate the price of your call or put option, by simply multiplying the option premium by the number I give you in each case. Once you get the hang of it, you'll be able to easily and quickly read any options table you want, for any of the major futures markets

I'll be covering 40 futures markets, including financials and stock indexes, including:

GRAINS: CBOT *Wheat, Corn, Oats, Soybeans, Soybean Oil, Soybean Meal, KBOT Wheat.*

MEATS: *Cattle, Feeder Cattle, Hogs, Pork Bellies.*

FOODS: *Sugar, Coffee, Cocoa, Orange Juice.*

PRECIOUS METALS: *CBOT Silver, Gold, CMX Silver, Platinum.*

INDUSTRIAL METALS: *Hi Grade Copper.*

OILS: *Light Sweet Crude, Heating Oil, Unleaded Gasoline, Natural Gas*

WOODS & FIBRES: *Lumber, Cotton.*

STOCK INDEXES: *S&P Comp., Nasdaq 100, Mini Value Line, NYSE Composite.*

FINANCIALS: *10 Yr. Treasury, US Treasury Bonds, Municipal Bonds, Eurodollars, British Pound, Canadian Dollar, German Mark, Japanese Yen, Swiss Franc, US Dollar Index.*

Before we begin, these are two abbreviations you'll need to be aware of:

no op - Stands for "no option." There's no option available for that particular strike price.

no tr - Stands for "no trade." Meaning that a particular strike price didn't trade that day. You'd be able to obtain more information from your broker for its current value.

GRAINS

```
Wheat - 5,000 bushels min. cents per bushel
  Prev day Call Vol  3,800  Open Int 70,757
  Prev day Put Vol   1,257  Open Int 38,103
(CBOT)
STRIKE       CALLS              PUTS
PRICE  DEC  MAR   MAY   DEC   MAR   MAY
370    72 1/2  no op  no op  no tr  no op  no op
380    66 1/2  no op  no op  no tr  no op  no op
390    44 3/4  44 1/2 no op  2 1/2  9 1/2  no op
400    no tr   38     no op  4 1/4  13     no op
410    29 1/2  32 3/4 no op  7 1/4  17     no op
420    23 1/4  27 1/4 25     11     21 3/4 no tr
430    (18 1/4) 23    21 1/2 15 1/2 27     no tr
440    14      19 1/4 no tr  21 1/4 33     no tr
450    10 3/4  16 1/4 no op  28     39 3/4 no op
460    8       13 3/4 no tr  35     47     no tr
```

CBOT Dec. Wheat Futures trading at 446 1/2.

Option Price = **Premium x $50.00**

Example:
18 1/4 x $50.00 = $912.50

COMMODITIES AND OPTIONS TRADING FOR BEGINNERS   153

```
Corn - 5,000 bushels min. cents per bushel
  Prev day Call Vol 11,100  Open Int 268,373
  Prev day Put Vol   7,822  Open Int 211,830
(CBOT)
STRIKE      CALLS              PUTS
PRICE  DEC   MAR   MAY    DEC   MAR   MAY
300    35 1/2 40 1/2 48 7/8  3 1/4  4 1/4  4 3/4
310    28 1/2 38 3/4 42 3/4  5 7/8  6 3/4  7
320    22 1/4 30 1/4 36 1/2  9 7/8  10 1/4 10 1/4
330    17     25     31      14 3/4 14 1/2 14 1/2
340     9 7/8 16     21 3/4  26 1/4 25 3/4 25
350     7 1/2 13 1/4 18 1/4  34 1/4 32 1/2 no tr
360    (5 3/8) 10 1/4 15 1/2 42 5/8 37 1/2 36 3/4
370     4 1/4  8 3/4 12 3/4  51     47 1/4 no tr
380     3      7     10 3/4  60     55 1/2 no tr
390     2 1/2  4 1/2  7 1/2  78 3/8 no tr  no tr
```

CBOT Dec. Corn Futures trading at 355.

Option Price = **Premium x $50.00**

Example:
5 3/8 x $50.00 = $268.75

GRAINS (cont.)

```
Oats - 5,000 bushels min. cents per bushel
  Prev day Call Vol 62  Open Int 3,453
  Prev day Put Vol  11  Open Int 1,574
(CBOT)
STRIKE      CALLS             PUTS
PRICE  DEC   MAR   MAY    DEC   MAR   MAY
150    34 1/4 no op no op  1 1/4  no op  no op
160    no tr  no op no op  5      no op  no op
170    12 1/4 no tr no op (9)     no tr  no op
180     7 3/4 12 1/2 no op 14 1/4 12 1/2 no op
190     5      9     no op 21 1/2 19 3/4 no op
200     2 3/4  6     no op 29 1/4 no tr  no op
```

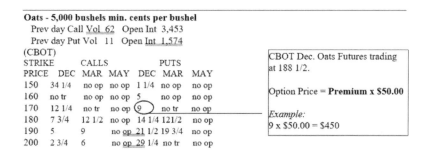

CBOT Dec. Oats Futures trading at 188 1/2.

Option Price = **Premium x $50.00**

Example:
9 x $50.00 = $450

Soybeans - 5,000 bushels min. cents per bushel
Prev day Call Vol 13,087 Open Int 154,191
Prev day Put Vol   8,513 Open Int 84,966
(CBOT)

STRIKE	CALLS			PUTS		
PRICE	DEC	MAR	MAY	DEC	MAR	MAY
700	100 3/4	109	113	1 1/4	3 1/4	4 1/2
725	77 1/2	87 1/2	94	3	6 1/2	9 1/2
750	56 1/2	69	76 1/2	7 1/8	13 1/2	17 1/8
775	39 1/4	54	62 1/2	15	23	27 1/4
800	29 1/2	42	50	27	35 1/2	39 3/4
825	18 3/4	33 1/2	40 1/2	43 1/4	51	no tr
850	12 5/8	25 3/8	32 3/4	64	68	no tr
875	8 1/4	19 1/4	26	82	no tr	89 1/2
900	5 3/4	15 1/4	20 3/4	105	no tr	108
925	4	11	16 1/2	no tr	no tr	no tr

> CBOT Soybeans Futures trading at 820.
>
> Option Price = **Premium x $50.00**
>
> *Example:*
> 29 1/2 x $50.00 = $1,475

GRAINS (cont.)

Soybean Oil -60,000 lbs., dollars per 100 lbs.
Prev day Call Vol 1,273 Open Int 22,261
Prev day Put Vol   102 Open Int 15,575
(CBOT)

STRIKE	CALLS			PUTS		
PRICE	DEC	MAR	MAY	DEC	MAR	MAY
235	no tr	no tr	no op	0.01	0.08	no op
240	1.71	2.29	no tr	0.02	0.13	0.15
245	1.28	1.87	no op	0.08	0.20	no op
250	0.86	1.53	no tr	0.15	0.34	0.40
255	0.50	1.21	no op	0.30	0.53	no op
260	0.28	0.98	1.25	0.60	0.79	0.80
265	0.15	0.79	no op	no tr	1.09	no op
270	0.07	0.49	0.70	no tr	1.79	no tr
275	0.05	0.45	0.55	no tr	2.17	no tr
280	no op	0.32	no tr	no op	no tr	no tr

> CBOT Soybean Oil Futures trading at 25.50.
>
> Option Price = **Premium x $6.00**
>
> *Example:*
> 50 x $6.00 = $300

Soybean Meal -100 tons, dollars per ton
Prev day Call Vol 935   Open Int 23,935
Prev day Put Vol 1,144 Open Int 14,717
(CBOT)

STRIKE PRICE	CALLS DEC	MAR	MAY	PUTS DEC	MAR	MAY
240	28.35	25.85	25.45	0.05	1.35	2.00
250	18.30	17.50	18.00	0.25	2.90	4.25
260	12.25	13.40	14.50	1.25	6.25	8.50
270	(2.50)	5.00	6.50	5.25	12.50	no tr
280	1.00	3.50	4.50	no tr	27.60	no tr
290	0.30	1.50	2.40	no tr	no tr	no tr

CBOT Soybean Meal Futures trading at 267.

Option Price = **Premium x $1.00**

Example:
250 x $1.00 = $250

GRAINS (cont.)

Wheat - 5,000 bushels min., cents per bushel
Prev day Call Vol 67 Open Int 4,611
Prev day Put Vol 66 Open Int 1,516
(KBOT)

STRIKE PRICE	CALLS DEC	MAR	MAY	PUTS DEC	MAR	MAY
410	33	no tr	no op	7 3/4	no tr	no op
420	26 1/4	no op	no tr	11 1/4	no op	no tr
430	20	no tr	no op	(14 3/4)	24	no op
440	16	23 1/2	no tr	21 1/4	30 1/2	no tr
450	13	20	no op	27 1/2	36 3/4	no op
460	10	16	no tr	34 1/4	42 1/2	no tr
470	7 1/2	14	13 3/4	42	50	no tr
480	5 1/2	11	no op	50	57 1/2	no op
490	4	9	no tr	58 1/4	66	no tr
500	2 3/4	7 1/2	no op	67	73 1/4	no tr

KBOT Wheat Futures trading at 450 1/2.

Option Price = **Premium x $50.00**

Example:
14 3/4 x $50.00 = $737.50

MEATS

Cattle -40,000 lbs., cents per lb.
Prev day Call Vol 810 Open Int 17,882
Prev day Put Vol 593 Open Int 22,275
(CME)

STRIKE PRICE	CALLS DEC	MAR	MAY	PUTS DEC	MAR	MAY
65	no op	7.22	4.12	no op	0.07	0.60
66	no op	6.25	3.25	no op	0.10	0.75
67	no op	5.27	2.57	no op	0.12	1.02
68	4.17	4.30	1.95	no op	0.22	no tr
69	no op	3.37	1.42	no op	0.35	2.35
70	no op	2.52	0.95	no op	0.60	no tr
71	no op	(1.77)	0.65	no op	0.92	no tr
72	no op	1.10	0.42	no tr	no tr	no tr

CME Cattle Futures trading at 68.57.

Option Price = **Premium x $4.00**

Example:
177 x 4.00 = $708

MEATS (Cont.)

Feeder Cattle - 50,000 lbs., cents per lb.
Prev day Call Vol 321 Open Int 5,856
Prev day Put Vol 183 Open Int 7,026
(CME)

STRIKE PRICE	CALLS DEC	MAR	MAY	PUTS DEC	MAR	MAY
61	no tr	no op	no op	0.05	no op	no op
62	2.10	2.85	3.35	0.12	0.60	0.72
63	no tr	no op	no tr	0.40	no op	no tr
64	0.67	1.52	2.00	(0.67)	1.25	no tr
65	0.30	no op	no tr	no tr	no op	no tr
66	0.12	0.70	1.10	no tr	2.40	no tr

CME Feeder Cattle Futures trading at 64.

Option Price = **Premium x $5.00**

Example:
67 x $5.00 = $335

COMMODITIES AND OPTIONS TRADING FOR BEGINNERS

Hogs (lean)- 40,000 lbs., cents per pound
Prev day Call Vol 243 Open Int 7,088
Prev day Put Vol 669 Open Int 9,836
(CME)

STRIKE PRICE	CALLS DEC	MAR	MAY	PUTS DEC	MAR	MAY
49	no tr	no tr	no op	0.17	0.45	no op
50	3.95	5.40	no op	0.20	0.55	no op
51	3.12	no tr	no op	0.35	0.87	no op
52	2.35	4.00	no op	0.55	1.15	no op
53	1.75	3.37	no op	0.97	1.50	no op
54	1.25	2.71	no op	1.47	1.90	no op
55	(0.85)	2.35	no op	2.02	no tr	no op
56	0.53	1.90	no tr	2.75	2.97	no tr
57	0.35	1.55	no op	no tr	no tr	no tr
58	0.20	1.22	no tr	4.40	4.27	no tr

CME Dec. Hogs Futures trading at 53.77.

Option Price = **Premium x $4.00**

Example:
85 x $4.00 = $340

MEATS (Cont.)

Pork Bellies -40,000 lbs., cents per lb.
Prev day Call Vol 38 Open Int 1,001
Prev day Put Vol 66 Open Int 1,908
(CME)

STRIKE PRICE	CALLS JUN	JUL	SEP	PUTS JUN	JUL	SEP
30	no op	no tr	no tr	no op	0.35	no tr
32	3.00	3.85	no op	0.50	1.50	no op
34	3.50	3.27	no op	1.52	1.97	no op
36	2.12	2.10	no op	2.20	4.00	no op
38	1.25	1.50	no op	2.95	4.55	no tr
40	1.00	1.15	no tr	4.65	no tr	no tr
42	(0.39)	0.80	no tr	5.82	6.75	no tr
44	0.27	0.44	no tr	7.75	no tr	no tr
46	0.15	0.37	no tr	9.00	no tr	no tr

CME June Pork Belly Futures trading at 38.57

Option Price = **Premium x $4.00**

Example:
39 x $4.00 = $156

FOODS (Cont.)

```
Sugar- World 11 -112,00 lbs., cents per lb.
    Prev day Call Vol  1,800  Open Int 75,061
    Prev day Put Vol   1,719  Open Int 52,521
(CSCE)
STRIKE      CALLS              PUTS
PRICE  OCT  MAR   MAY    OCT   MAR   MAY
950    2.51 2.23  2.02   0.01  0.04  0.09
1000   2.01 1.73  1.54   0.01  0.07  0.16
1050   1.51 1.22  1.20   0.01  0.13  0.27
1100   1.00 0.95  0.83   0.01  0.25  0.42
1150   0.34 0.62  0.57   0.03  0.54  0.63
1200  (0.13)0.39  0.40   0.13  0.70  0.93
1250   0.02 0.25  0.24   0.53  1.06  1.31
1300   0.01 0.15  0.17   1.01  1.40  1.71
1350   0.01 0.08  0.11   2.01  2.36  2.65
1400   0.01 0.04  no op  2.51  2.85  no op
```

CSCE Oct. Sugar Futures trading at 12.00.

Option Price = **Premium x $11.20**

Example:
13 x $11.20 = $145.60

FOODS (Cont.)

COMMODITIES AND OPTIONS TRADING FOR BEGINNERS

Coffee C - 37,500 lbs., cents per lb.
Prev day Call Vol 898   Open Int 30,127
Prev day Put Vol 840   Open Int 15,917
(CSCE)

STRIKE PRICE	CALLS OCT	DEC	MAR	PUTS OCT	DEC	MAR
1000	12.90	14.70	13.50	0.02	1.80	5.10
1050	7.90	11.15	10.60	0.10	3.25	7.20
1100	(2.90)	8.00	8.20	0.01	5.10	9.80
1150	2.02	5.20	6.35	2.10	7.80	13.00
1200	0.02	3.70	5.00	7.10	10.80	16.80
1250	no tr	2.56	3.80	12.00	14	20.40
1300	no tr	1.70	2.90	17.10	18.88	24.40
1350	no tr	1.25	2.10	22.10	23.35	28.70
1400	no tr	0.90	1.60	27.10	28.00	33.20
1450	no tr	0.70	1.25	32.10	32	37.85

CSCE Oct. Coffee Futures trading at 124.60.

Option Price = **Premium x $3.75**

Example:
290 x $3.75 = $1,087.50

Orange Juice - 15,000 lbs., cents per lb.
Prev day Call Vol 1,069   Open Int 24,116
Prev day Put Vol   662   Open Int 22,854
(CTN)

STRIKE PRICE	CALLS OCT	NOV	JAN	PUTS OCT	NOV	JAN
1000	no tr	7.20	9.90	no tr	1.40	4.75
1050	no tr	(4.00)	7.10	no tr	3.00	6.85
1100	no tr	2.10	5.05	no tr	9.90	13.10
1150	no tr	1.05	3.50	no tr	14.40	16.85
1200	no tr	0.50	2.75	no tr	19.20	21.00
1250	no tr	0.25	2.00	no tr	24.00	25.35

CTN Oct. Orange Juice Futures trading at 115.75.

Option Price = **Premium x $1.50**

Example:
400 x $1.50 = $600

FOODS (Cont.)

Cocoa - 10 metric tons, dollars per ton
Prev day Call Vol 429   Open Int 20,860
Prev day Put Vol 539   Open Int 9,501
(CSCE)

STRIKE PRICE	CALLS			PUTS		
	OCT	DEC	MAR	OCT	DEC	MAR
1150	223	224	263	no tr	1	6
1200	173	175	276	no tr	2	9
1250	123	129	173	no tr	6	16
1300	73	85	131	1	12	24
1350	23	53	100	1	30	43
1400	2	34	75	27	58	69
1450	no tr	(20)	59	77	97	102
1500	no tr	11	43	127	138	136
1550	no tr	7	33	177	184	176
1600	no tr	4	28	227	231	221

CSCE Oct. Cocoa Futures trading at 13.50.

Option Price = **Premium x $10.00**

Example:
20 x $10.00 = $200

PRECIOUS METALS

Silver (1000)- 1,000 oz, cents per oz.
Prev day Call Vol 2 Open Int 268
Prev day Put Vol 1 Open Int 1
(CBOT)

STRIKE PRICE	CALLS			PUTS		
	OCT	DEC	APR	OCT	DEC	APR
525	no tr	no op	100	4	no op	20
550	55	no op	no op	11	45	no op
575	(45)	no op	no op	8	13	no op
600	3	no op	no op	15	30	no op
625	35	40	25	no tr	no tr	no op
650	31	15	45	no op	no op	no op
675	17	5	8	no op	no op	no op
700	no op	no op	42	no op	no op	no op

CBOT Oct. Silver Futures trading at 597.

Option Price = **Premium x $1.00**

Example:
45 x 1.00 = $45

PRECIOUS METALS (Cont.)

Gold - 100 troy ounce, dollar per troy ounce
Prev day Call Vol 10,900 Open Int 251,486
Prev day Put Vol 5,592 Open Int 78,739
(CMX)

STRIKE PRICE	CALLS OCT	NOV	DEC	PUTS OCT	NOV	DEC
350	42	45.50	46.80	0.10	0.10	0.20
360	26.50	27.80	28.90	0.10	0.10	0.30
370	17.00	22.50	25.70	no tr	0.40	0.50
380	11.00	14.00	(14.70)	0.20	0.80	1.20
390	0.10	1.80	4.80	5.50	5.70	7.50
400	0.10	0.90	2.30	12.70	13.80	13.10

CMX Oct. Gold Futures trading at 386.70

Option Price = **Premium x $1.00**

Example:
1470 x $1.00 = $1,470

Silver - 5,000 troy oz., cents per troy ounce
Prev day Call Vol 1,574 Open Int 60,244
Prev day Put Vol 684 Open Int 18,872
(CMX)

STRIKE PRICE	CALLS OCT	NOV	DEC	PUTS OCT	NOV	DEC
400	115.5	115.8	115.8	0.1	0.1	0.1
425	90.8	90.8	90.8	0.1	0.1	0.1
450	65.8	65.8	65.8	0.1	0.1	0.1
475	40.8	41.0	42.0	0.2	0.8	2.2
500	16.8	19.8	22.5	1.0	4.0	7.0
525	1.8	6.6	(8.2)	11.0	15.8	19.4
550	0.4	2.3	4.8	34.8	36.5	39.3
575	0.2	1.0	2.5	58.4	60.0	61.5
600	0.1	0.5	1.7	84.2	no tr	85.5
625	0.1	0.3	1.2	111.2	109.5	112.5

CMX Oct. Silver Futures trading at 510.50.

Option Price = **Premium x $5.00**

Example:
82 x $5.00 = $410

PRECIOUS METALS (Cont.)

```
Platinum - 50 troy oz., dollars per troy oz.
    Prev day Call Vol  5   Open Int  2,836
    Prev day Put Vol   4   Open Int    482
(CMX)
STRIKE         CALLS              PUTS
PRICE  OCT   JAN   APR    OCT   JAN   APR
 370   26.00 29.50 no tr  no tr no tr no tr
 380   16.00 20.00 no tr  no tr no tr no tr
 390    6.20 12.50 no tr   0.30  4.00 no tr
 400    0.60  6.80 no tr   4.70  8.30 no tr
 410    0.10  3.50 no tr  14.70 15.00 no tr
 420    0.10  2.30  6.50  no tr 21.50 no tr
 430    0.10  no op no tr no tr no tr no tr
 440    0.10  no tr no tr no tr no tr no tr
 450    0.10  no tr no tr no tr no op no tr
```

CMX Oct. Platinum Futures trading at 395.90.

Option Price = **Premium x $5.00**

Example:
62.0 x $5.00 = $310

INDUSTRIAL METALS

INDUSTRIAL METALS

```
Hi Grade Copper - 25,000 lbs., cents per lb.
    Prev day Call Vol   313   Open Int  13,721
    Prev day Put Vol    162   Open Int   2,165
(CMX)
STRIKE         CALLS              PUTS
PRICE  OCT  NOV   DEC    OCT   NOV   DEC
  82   8.10 8.80  9.65   0.30  1.0   1.85
  84   6.40 7.30  8.45   0.50  1.50  2.45
  86   4.70 6.00  7.00   0.90  2.20  3.20
  88   3.30 4.80  5.90   1.30  3.0   4.30
  90   2.20 3.60  4.90   2.40  3.40  5.10
  92   1.30 2.80  4.10   3.50  4.40  6.30
  94   0.80 2.20  3.30   5.00  6.40  7.50
  96   0.55 1.60  2.70   6.70  7.80  8.30
  98   0.40 1.30  2.20   8.60  9.40 10.40
 100   0.30 0.90  1.70  10.50 11.10 11.90
```

CMX Oct. Copper Futures trading at 90.45.

Option Price = **Premium x $2.50**

Example:
280 x $2.50 = $700

OILS

Light Sweet Crude - 1,000 bbl., $ per bbl.
Prev day Call Vol 16,843 Open Int 261,304
Prev day Put Vol 9,506 Open Int 181,042
(NYM)

STRIKE PRICE	CALLS OCT	NOV	DEC	PUTS OCT	NOV	DEC
2150	2.87	2.50	2.24	0.02	0.34	0.66
2200	2.38	2.13	1.91	0.03	0.47	0.83
2250	1.90	1.81	1.63	0.05	0.64	1.04
2300	1.44	1.51	1.39	0.09	0.84	1.29
2350	(1.05)	1.24	1.18	0.20	1.07	1.59
2400	0.40	0.94	1.00	0.57	1.68	no tr
2450	0.12	0.60	0.61	1.31	2.42	no tr
2500	0.10	no op	no op	no op	no op	no op
2550	0.10	no op	no op	no tr	no tr	no op

NYM Oct. Crude Oil Futures trading at 23.85.

Option Price = **Premium x $10.00**

Example:
105 x $10.00 = $1,050

Natural Gas -10,000 mm btu's, $ per mm btu.
Prev day Call Vol 3,349 Open Int 75,479
Prev day Put Vol 794 Open Int 51,565
(NYM)

STRIKE PRICE	CALLS OCT	NOV	DEC	PUTS OCT	NOV	DEC
165	no tr	no op	no op	1.4	no tr	1.1
170	no tr	no tr	no tr	2.3	1.7	no tr
175	14.8	no tr	no tr	3.5	2.4	2.2
180	(11.8)	29.0	40.9	5.5	3.4	1.7
185	8.8	25.5	no tr	7.5	4.6	no tr
190	5.3	21.8	no tr	10.0	6.2	5.2

NYM Oct. Natural Gas Futures trading at 1,853.

Option Price = **Premium x $10.00**

Example:
118 x $4.20 = $495.60

OILS (Cont.)

Unleaded Gasoline - 42,000 gal., cents per gal.
Prev day Call Vol 513 Open Int 11,986
Prev day Put Vol 687 Open Int 8,625
(NYM)

STRIKE PRICE	CALLS OCT	NOV	DEC	PUTS OCT	NOV	DEC
59	6.43	5.40	no op	0.17	0.95	no op
60	5.46	4.70	4.10	0.20	1.25	2.10
61	4.81	3.50	no op	0.40	1.75	no op
62	4.61	3.30	no tr	0.55	1.90	no op
63	3.81	3.21	2.87	0.50	no op	no op
64	2.42	2.49	2.25	1.15	2.98	no tr
65	1.80	no op	no op	no op	no op	no op
66	1.25	1.70	no op	no tr	no tr	no op
67	0.70	1.20	1.15	no tr	no tr	no tr
68	0.35	no op	no tr	no op	no tr	no tr

> NYM Oct. Unleaded Gasoline Futures trading 65.27.
>
> Option Price = **Premium x $4.20**
>
> *Example:*
> 55x $4.20 = $231

Heating Oil - 42,000 gal., cents per gallon
Prev day Call Vol 2,397 Open Int 67,135
Prev day Put Vol 1,322 Open Int 25,907
(NYM)

STRIKE PRICE	CALLS OCT	NOV	DEC	PUTS OCT	NOV	DEC
61	5.69	6.44	no op	0.30	1.15	no op
62	4.86	5.80	6.14	0.47	1.50	2.15
63	4.04	5.10	no tr	0.68	2.00	no op
64	3.35	4.52	5.01	0.95	2.20	3.00
65	2.70	3.97	no op	1.30	2.65	no op
66	2.15	3.47	4.10	1.75	3.15	4.15
67	1.35	2.80	3.35	no tr	4.45	no tr
68	0.75	no op	no op	no tr	no op	no op
69	0.50	2.10	2.75	no tr	no tr	no tr
70	no op	1.30	2.25	no tr	no tr	2.15

> NYM Oct. Dec. Heating Oil Futures trading at 66.40.
>
> Option Price = **Premium x $4.20**
>
> *Example:*
> 135 x $4.20 = $567

WOODS & FIBRES

Lumber -160,000 bd. ft., $ per 1,000 bd. ft.
Prev day Call Vol 118 Open Int 788
Prev day Put Vol 63 Open Int 311
(CME)

STRIKE PRICE	CALLS NOV	JAN	MAR	PUTS NOV	JAN	MAR
355	32.80	no tr	no op	10.00	no op	no op
360	39.80	no op	no tr	7.00	no op	no op
365	32.80	27.30	no tr	8.50	no tr	no op
370	29.70	22.70	no op	11.80	no tr	no op
375	26.80	no op	no op	13.90	no op	no op
380	23.50	18.60	no op	16.00	no op	no op
385	19.00	15.10	no tr	21.00	no op	no op
390	14.80	no tr	no tr	26.70	no tr	no tr
395	11.40	no op	no op	no tr	no op	no op
400	8.50	no op	no op	no tr	no op	no op

CME Nov. Lumber Futures trading at 388.00.

Option Price = **Premium x $1.60**

Example:
2350 x $4.00 = $9,400

Cotton 2 - 50,000 lbs., cents per lb.
Prev day Call Vol 1,990 Open Int 56,205
Prev day Put Vol 1,700 Open Int 46,111
(CTN)

STRIKE PRICE	CALLS OCT	DEC	MAR	PUTS OCT	DEC	MAR
69	no op	no tr	no tr	0.17	0.99	1.84
70	3.43	5.37	7.41	0.26	1.25	2.20
71	2.63	4.71	6.78	0.44	1.57	2.49
72	1.93	4.09	6.19	0.77	1.94	2.87
73	1.35	3.62	5.83	1.20	(2.26)	3.50
74	0.57	2.56	4.62	2.43	3.37	4.21
75	0.34	2.16	4.17	3.30	3.95	4.73
76	0.10	1.59	3.75	4.07	4.59	5.28
77	0.07	1.24	3.00	5.94	5.98	6.49

> CTN Oct. Cotton Futures trading at 73.10.
>
> Option Price = **Premium x $5.00**
>
> *Example:*
> 226 x $5.00 = $1,130

STOCK INDEX

S&P Comp Index - 500 x Premium
Prev day Call Vol 5,103 Open Int 96,742
Prev day Put Vol 13,688 Open Int 143,648
(CME)

STRIKE PRICE	CALLS SEP	DEC	MAR	PUTS SEP	DEC	MAR
630	29.55	49.75	no tr	1.35	11.00	16.20
635	29.90	no tr	no tr	1.70	12.15	no tr
640	20.45	37.30	no tr	2.25	13.15	no tr
645	16.25	33.75	no tr	3.05	14.80	no tr
650	12.40	30.35	41.70	4.15	16.30	21.00
655	9.00	27.05	no tr	5.75	17.60	no tr
660	(6.10)	24.80	35.30	7.85	19.80	33.00
665	3.20	21.00	32.70	10.60	21.05	no tr
670	2.30	18.45	20.20	14.00	24.60	29.30
675	1.30	15.85	25.80	16.80	29.50	31.20

> CME Sep. S&P Futures trading at 655.30.
>
> Option Price = **Premium x $5.00**
>
> *Example:*
> 610 x $5.00 = $3,050

COMMODITIES AND OPTIONS TRADING FOR BEGINNERS 167

```
NASDAQ 100 Index - $100 x index.
   Prev day Call Vol 22   Open Int 241
   Prev day Put Vol  52   Open Int 177
(CME)
STRIKE        CALLS            PUTS
PRICE  SEP    DEC          SEP    DEC
6100   no tr  no tr  no op  0.05   no tr  no op
6200   no tr  no tr  no op  0.30   no tr  no op
6300   no tr  no tr  no op  1.50   no tr  no op
6400   29.50  no tr  no op  3.80   19.75  no op
6500   21.70  no op  no op  6.00   no op  no op
6600   15.0   no op  no op  9.25   no op  no op
6700   (9.20) no op  no op  13.50  no op  no op
6800   5.00   no op  no op  19.20  no op  no op
6900   2.40   22.95  no op  26.60  no tr  no op
7000   1.10   no op  no op  35.30  no op  no op
```

CME NASDAQ Futures trading at 666.75

Option Price = **Premium x $1.00**

Example:
920 x $1.00 = $920

STOCK INDEX (Cont.)

```
Municipal Bonds 1000 x index, pts & 64ths of 100
   Prev day Call Vol 20   Open Int 1,275
   Prev day Put Vol   3   Open Int 3,321
(CBOT)
STRIKE        CALLS            PUTS
PRICE  SEP    DEC          SEP    DEC
108    no tr  no op  no op  1/64   no op  no op
109    no tr  no op  no op  2/64   no op  no op
110    no tr  no op  no op  3/64   no tr  no op
111    no tr  (1 2/64) no op 7/64  no tr  no op
112    no tr  no tr  no op  14/64  no tr  no op
113    no op  no op  no op  50/64  no tr  no op
```

CBOT Sep. Municipal Bonds Futures trading at 111.14.

Option Price = **Premium x $15,625, PLUS full points x $1,000**

Example:
2 x $15.625 = $31.25 PLUS
2 x $1,000 = $2,031.25

NYSE Comp Index - 500 x Premium
  Prev day Call Vol 96 Open Int 1,319
  Prev day Put Vol 137 Open Int 2,479
(NYFE)

STRIKE PRICE	CALLS SEP	DEC	MAR	PUTS SEP	DEC	MAR
338	14.50	22.75	28.75	0.35	5.25	8.35
340	12.90	20.75	26.00	0.75	5.65	8.75
342	11.70	20.00	25.20	1.15	7.00	9.10
344	10.90	19.50	25.00	1.45	7.85	10.35
346	6.85	18.30	24.05	1.80	no tr	10.95
348	no tr	16.55	22.65	2.30	no tr	12.00
350	(6.05)	15.50	21.30	2.80	9.80	12.25
352	4.20	14.20	19.70	3.60	11.05	13.00
354	3.30	13.45	no tr	4.50	11.55	13.85
356	2.00	11.75	no tr	5.90	12.50	14.40

September NYSE Composite Futures trading at 349.15.

Option Price = **Premium x $5.00**

Example:
605x $5.00 = $3,025

FINANCIALS

10 Yr. Treasury -$100,000 prin., pts & 64ths of 100 pct.
Prev day Call Vol 22,400   Open Int 141,118
Prev day Put Vol 29,346   Open Int 119,072
(CBOT)

STRIKE	CALLS			PUTS		
PRICE	OCT	DEC	MAR	OCT	DEC	MAR
100	no tr	6 57/64	7	1/64	12/64	46/64
101	no tr	no tr	no op	1/64	no tr	no op
102	no tr	5 8/64	no op	1/64	no tr	no op
103	no tr	no tr	no op	2/64	no tr	no op
104	2 45/64	2 16/64	no op	6.00	5 2/64	1 50/64
105	1 63/64	no op	no op	15/64	no op	no op
106	1 16/64	14/64	no op	130/64	2 38/64	no op
107	44/64	47/64	no op	2 24/64	no op	no op
108	22/64	118/64	2 3/64	3 19/64	no tr	no op
109	8/64	60/64	no op	4 17/64	no tr	no op

October 10-Year Treasury Bond Futures trading at 105.10

Option Price = **Premium x $15.625, then add full points x $1,000.**
Example:
2 x $15.625 = $31.25

Example:
30 x $15.625 = $468.75 PLUS
1 x $1,000 = $1,468.75

Mini Value Line - points and cents
Prev day Call Vol 2   Open Int 18
Prev day Put Vol 3   Open Int 21
(KBOT)

STRIKE	CALLS			PUTS		
PRICE	MAY	JUN	SEP	MAY	JUN	SEP
107	no op	no op	no op	1/64	no op	no op
108	2 24/64	no op	no op	2/64	no op	no op
109	no op	2 3/64	no op	7/64	no tr	no op
110	no tr	no tr	no op	14/64	no tr	no op
111	8/64	no tr	no op	1 36/64	no tr	no op
112	3/64	no tr	no op	no tr	no tr	no op

May Municipal Bond Futures trading at 109.15.

Option Price = **Premium x $15.625, then add full points x $1,000.**
Example:
24 x $15.625 = $375, PLUS
2 x $1,000 = $2,375

FINANCIALS (Cont.)

US Treasury Bonds – (8 pct=$100,000 pts & 64ths of 100 %)
Prev day Call Vol 85,400 Open Int 336,744
Prev day Put Vol 64,037 Open Int 224,008
(CBOT)

STRIKE PRICE	CALLS OCT	DEC	MAR	PUTS OCT	DEC	MAR
101	no tr	no tr	no op	1/64	no op	no op
102	no tr	5 8/64	no op	1/64	27/64	1 12/64
103	no tr	no op	no tr	2/64	no tr	no tr
104	2 4/64	3 35/64	no tr	6/64	52/64	1 51/64
105	1 63/64	no tr	no tr	15/64	1 8/64	no op
106	1 16/64	2 6/64	2 62/64	32/64	1 32/64	2 32/64
107	44/64	1 8/64	2 3/64	60/64	no op	no op
108	8/64	60/64	no op	2 24/64	no tr	no tr
109	3/64	42/64	no op	3 19/64	3 56/64	4 57/64
110	1/64	28/64	no op	4 17/64	5 33/64	6 22/64

Oct. Treasury Bond Futures trading at 107.12

Option Price = **Premium x $15.625, PLUS full points x $1,000**

Example:
19 x $15,625 = PLUS 3 x $1,000 = $3,296.87

Eurodollars – $1 million, pts of 100 pct.
Prev day Call Vol 61,083 Open Int 1,872,444
Prev day Put Vol 88,997 Open Int 1,196,159
(CME)

STRIKE PRICE	CALLS SEP	OCT	NOV	PUTS SEP	OCT	NOV
9300	1.30	no tr	no op	no tr	no tr	no op
9325	1.05	no tr	no op	0.01	0.02	no op
9350	0.80	no tr	no op	no tr	0.02	0.03
9375	0.55	0.23	no tr	no tr	0.06	0.08
9400	0.30	0.08	0.11	no tr	0.16	0.19
9425	0.07	0.02	0.03	0.02	0.35	0.36
9450	0.01	no tr	no tr	0.08	no op	no tr
9475	no tr	no tr	no op	0.45	no tr	no tr
9500	no tr	no tr	no op	0.70	no tr	no tr
9525	no tr	no tr	no op	0.95	no tr	no tr

Sept. Eurodollar Futures trading at 94.30

Option Price = **Premium x $25.00**

Example:
30 x $25.00 = $750

FINANCIALS (Cont.)

British Pound - cents per pound
Prev day Call Vol 3,113   Open Int 1,275
Prev day Put Vol   859   Open Int 3,321
(CME)

STRIKE PRICE	CALLS SEP	DEC	MAR	PUTS SEP	DEC	MAR
1530	3.02	no tr	no tr	no tr	no tr	no op
1540	2.02	2.28	3.96	no tr	0.90	1.74
1550	1.02	no tr	no op	no tr	no tr	no op
1560	0.04	1.72	2.76	0.02	(1.72)	2.58
1570	no tr	no tr	no op	0.98	no tr	no op
1580	no tr	0.90	1.88	1.98	2.88	no tr

Sep. British Pound Futures trading at 1.560.

Option Price = **Premium x $6.25**

Example:
172 x $6.25 = $1,075

Canadian Dollar - cents per dollar
Prev day Call Vol 126   Open Int 16,200
Prev day Put Vol 229   Open Int 9,093
(CME)

STRIKE PRICE	CALLS SEP	DEC	MAR	PUTS SEP	DEC	MAR
705	2.88	no tr	no op	no tr	0.01	no tr
710	1.89	no tr	no tr	no tr	0.04	no tr
715	no tr	no tr	no tr	no tr	0.08	0.17
720	0.88	1.28	1.62	no tr	0.15	0.26
725	0.38	0.90	1.26	no tr	0.27	0.39
730	no tr	0.61	0.95	0.12	0.47	0.57
735	no tr	0.38	0.68	(0.62)	0.74	0.89
740	no tr	0.23	0.48	1.12	1.08	1.10
745	no tr	0.12	0.32	no tr	1.46	no tr
750	no tr	0.07	0.21	no tr	1.91	no tr

Sep. Canadian Dollar Futures trading at .7288

Option Price = **Premium x $10.00**

Example:
62 x $10.00 = $620

FINANCIALS (Cont.)

German Mark - cents per mark
 Prev day Call Vol 3,277 Open Int 62,483
 Prev day Put Vol 2,659 Open Int 54,150
(CME)

STRIKE PRICE	CALLS SEP	DEC	MAR	PUTS SEP	DEC	MAR
645	2.54	no tr	no op	no tr	no tr	no op
650	2.04	2.67	3.39	no tr	0.23	0.50
655	1.54	no tr	no tr	no tr	no tr	no op
660	1.04	1.89	2.57	no tr	0.44	0.76
665	(0.54)	no tr	no op	0.01	no tr	no op
670	0.05	1.25	2.06	0.01	0.79	1.12
675	no tr	no tr	no op	0.46	no tr	no tr
680	no tr	0.77	1.54	0.96	1.29	1.58
685	no tr	no tr	no op	1.46	no tr	no tr
690	no tr	0.47	1.13	1.96	1.98	2.14

Sep. German MarkFutures trading at .6704

Option Price = **Premium x $12.50**

Example:
54 x $12.50 = $675

Japanese Yen - cents per 100 yen
 Prev day Call Vol 1,976 Open Int 59,179
 Prev day Put Vol 3,445 Open Int 39,247
(CME)

STRIKE PRICE	CALLS SEP	DEC	MAR	PUTS SEP	DEC	MAR
900	2.48	3.90	no tr	no tr	0.26	0.55
905	1.48	3.09	4.60	no tr	0.44	0.76
910	0.98	no op	no op	0.01	no tr	no tr
915	0.04	no op	no op	0.05	no op	no op
920	no tr	1.74	3.26	(0.52)	1.07	1.37
925	no tr	no tr	no op	1.02	no tr	no op

Sep. Japanese Yen Futures trading at .009148

Option Price = **Premium x $12.50**

Example:
52 x $12.50 = $650.00

FINANCIALS (Cont.)

Swiss Franc - cents per franc
Prev day Call Vol 1,500 Open Int 26,840
Prev day Put Vol 1,759 Open Int 22,237
(CME)

STRIKE PRICE	CALLS SEP	DEC	MAR	PUTS SEP	DEC	MAR
795	2.58	no tr	no op	no tr	no op	no op
800	2.05	3.24	no tr	no tr	0.49	0.89
805	1.58	no tr	no op	no tr	no op	no op
810	1.08	2.52	3.66	no tr	0.75	1.17
815	0.58	no op	no op	0.01	no op	no op
820	0.08	1.90	3.94	0.01	0.12	1.53
825	0.26	no op	no op	0.61	no op	no op
830	0.02	1.38	2.29	0.92	1.59	1.95
835	no tr	no op	no op	1.42	no op	no tr
840	no tr	0.98	2.04	1.92	2.17	2.48

Sep. Swiss Franc Futures trading at .8208

Option Price = **Premium x $12.50**

Example:
61 x $12.50 = $822.50

US Dollar Index – 1,000 x premium
Prev day Call Vol 36 Open Int 2,086
Prev day Put Vol 28 Open Int 1,627
(CTN)

STRIKE PRICE	CALLS SEP	DEC	JUN	PUTS SEP	DEC	JUN
81	no tr	no tr	no op	0.01	0.01	no tr
82	no tr	no tr	no tr	0.01	0.03	no tr
83	no tr	no tr	no tr	0.01	0.08	no tr
84	2.84	no tr	no tr	0.01	0.20	no tr
85	1.84	1.98	no tr	0.01	0.40	no tr
86	0.84	1.33	no tr	0.01	0.73	no tr
87	0.01	0.82	no tr	0.16	1.22	no tr
88	0.01	0.47	no tr	1.16	1.61	no tr
89	0.01	0.25	no tr	2.16	2.80	no tr
90	0.01	0.02	no tr	no tr	no tr	no tr

Sep. US Dollar Index Futures trading at 86.84

Option Price = **Premium x $10.00**

Example:
47 x $10.00 = $470.00

Chapter Fourteen

Conclusion

I've been telling you all along that by the time you're through with this book you'll know more about trading options than most of the people that currently trade them. I've done my best to deliver on this promise. You now have not only a solid foundation on options, but also something essential for success: peripheral vision.

Peripheral vision is what you need to avoid blind spots when you trade. Entering a trade without knowing the probabilities of success or failure, and without knowing just how much you stand to lose, is the quickest way to get in trouble.

But you've seen some of the main strategies to apply for many different situations.

You've learned how to assess the risks, the rewards, and the probabilities, and the importance of using the proper strategy to meet different risk/reward objectives.

We've examined the use of **spreads.** They can increase the probability of profits while reducing the risks, in case the price of the commodity moves beyond what you thought it would. We've seen how we can use combinations of different spread techniques, including box spreads, which allow you to profit in both directions.

We've also learned about **straddles** and **strangles**. We've seen how a long straddle creates the potential for profits with large price increases or decreases, while also creating a middle zone loss.

On the short straddle, we created a middle profit zone, while leaving open the possibilities of losses with extreme price moves in either direction. We also saw that by combining long and short strangles, we limited our potential gains, but also considerably limited our exposure to only smaller losses.

Finally, I've shown some examples of **hedging**.

By hedging, we were able to offset any potential losses from one position with potential gains from the other positions. You may also hedge to protect a position you've acquired in the futures market (or a particular stock from your stock portfolio.)

For instance, buying a put to protect against short sale losses. Again, one way or another, we're putting the odds at our favor.

Remember, while this book concentrated on options on futures, the principles and concepts can be easily transferred to the stock market or the stock index. The requirements are the same.

Let's recap what these requirements are:

1. With every transaction there'll be typically brokerage fees. You must figure them in to calculate your break-even point. These fees will vary from broker to broker.

2. Whenever you assume a short position, your broker will have margin requirements. You'll need to show and maintain some money (or securities) as collateral in your brokerage account. The amount required varies from broker to broker.

3. Before investing money, carefully go over the risk/reward factors involved with the strategy you're using. If you're not sure, then just 'paper trade' the first few times.

In other words, follow the option you would have been trading through the newspaper for a few weeks. This is going to help you get comfortable with the strategy you selected.

4. Set specific target goals. This is not really a requirement, but it's the right thing to do. If you enter a trade with a
specific dollar amount gain in mind, then you can close your position quickly when you reach that target.

Likewise, know exactly how much money you're willing to lose on the option. If you reach that figure, close your position quickly. It should be a methodical process.

Deciding on trading an option is a more complex decision then deciding on a brand of soap.

Granted, it's one thing to simply 'talk' about risking your money, but its' another thing to actually 'do it'.

However, keep in mind that you now have gotten the knowledge it takes.

At this point, you know how to make money with options. All you need is to build up your confidence level. Confidence will help you use this knowledge wisely. Don't enter a trade and let your emotions completely take over, throwing everything you've learned right out the window.

This book just demonstrated to you how to put the odds overwhelmingly on your side. Without using these techniques, you'll either be aligning yourself with option losers, or exposing yourself to potentially unlimited risks.

Instead, you can limit the maximum amount of money you can possibly lose to a specific figure. Not a penny more! This will certainly help you sleep better at night, specially if you're a risk-adverse type of person such as I am.

Good luck trading, and keep in control!

Glossary of Terms

Actuals: The physical or cash commodity, as opposed to the futures contract.

Aggregation: The policy under which all futures positions owned or controlled by one trader or a group of traders are combined to determine reporting status and speculative compliance.

Arbitrage: The simultaneous purchase and sale of the same or similar commodities in different markets in order to make a profit from the price discrepancy.

Arbitration: The process of settling disputes between members or between members and customers. NFA's arbitration program provides a forum for resolving futures-related disputes.

Associated Person (AP): An individual who solicits orders, customers or customer funds on behalf of a Futures Commission Merchant, an Introducing Broker, a Commodity Pool Operator or a Commodity Trading Advisor who is registered with the Commodity Futures Trading Commission.

At-the-Market: See Market Order

At-the-Money: An option whose strike price is equal - or approximately equal - to the current market price of the underlying futures contract.

Backwardation: A market in which futures prices are progressively lower in the distant delivery months; the opposite of Contango. See also Inverted Market.

Basis: The difference between the cash or spot price and the futures price of the same commodity.

Bear Market (Bear/Bearish) : A market in which prices are in decline. A market participant who believes prices will move lower is called a bear. A news item is considered bearish if it is expected to produce lower prices.

Bid: An offer to buy a commodity at a stated price; the opposite of offer.

Board of Trade: See Contract Market.

Break: A rapid and sharp price decline.

Broker: A person paid a fee or commission for acting as an agent in making contracts, sales or purchases. In futures trading, the term may refer to a floor broker -a person who actually executes orders on the trading floor of an exchange; or an Account Executive or Associated Person -who deals with customers in the offices of a Futures Commission Merchant or Introducing Broker; or a Futures Commission Merchant or Introducing Broker.

Brokerage: A fee charged by a broker for execution of a transaction. Commonly referred to as a commission fee.

Bucketing: Directly or indirectly taking the opposite side of a customer's order into the broker's own account or into an account in which the broker has an interest, without the open and competitive execution of the order on an exchange.

Bull Market (Bull/Bullish): A market in which prices are on the rise. A participant in futures who believes prices will move higher is called a bull. A news item is considered bullish if it is expected to bring higher prices.

Buying Hedge: Buying futures contracts to protect against the possible increase cost of commodities intended for future uses. Also see Long Hedge.

Call Option: The buyer of a call option acquires the right but not the obligation to purchase a particular futures contract at a stated price on or before a particular date.

Car(s): This is a colloquialism for futures contract(s). The term came into common use when a railroad car or hopper of a grain equaled the amount of a commodity in a futures contract.

Carrying Broker: A member of a futures exchange, usually a clearinghouse member, through whom a customer or another broker chooses to clear all or some trades.

Carrying Charges: The cost of storing a physical commodity, such as grain or metals, over a period of time. These costs include interest on the invested funds, insurance, storage and other incidental costs.

Carryover: The part of the current supply of a commodity consisting of stocks from previous production season.

Cash Commodity: The actual physical commodity as distinguished from the futures contract which is based on the physical commodity. Also known as Actuals.

Cash Market: A place where people buy and sell the actual commodities. See also Forward Contract and Spot.

Cash Settlement: A method of selling futures or options contracts where the seller pays the buyer the cash value of the commodity traded according to a procedure specified in the contract.

Certified Stock: Stocks of a commodity that has been inspected and found to be of a deliverable quality against a futures contract.

Charting: The use of graphs and charts in the technical analysis of futures markets to plot trends of price movements, volume, open interest or other statistical indicators of price movement. See also Technical Analysis.

Churning: Excessive trading of a customer's account by a broker who controls the decision making on the account, to make more commissions while ignoring the best interest of his customer.

Circuit Breaker: A system of trading halts and price limits on equities and derivatives markets designed to provide a cooling-off period during large, intraday market declines.

Clearing: The procedure through which trades are checked for accuracy. Once the trades are validated the clearing house becomes the buyer to each seller of a futures contract, and a seller to each buyer.

Clearing House: An agency or a fully charted separate corporation of a futures exchange through which financial settlement occurs. It is responsible for settling trading accounts, collecting and maintaining margin money, regulating delivery and reporting trade data.

Clearing Member: A member of an exchange clearing house. All trades of a non-clearing member must be registered and eventually settled through a clearing member.

Clearing Price: See Settlement Price

Close (the): The period at the end of the trading session, officially designated by the exchange, where all transactions are considered made "at the close."

Closing Range: The range of high and low prices at which futures transactions took place during the close of the market.

Commission: A fee charged by a broker to a customer for performance of a specific duty, such as the buying or selling of futures contracts.

Commission Merchant: One who makes a trade in his own name, for either another member of the exchange or for a non-member client, assuming all liabilities.

Commodity: An entity of trade or commerce, services, or rights upon which contracts for future delivery may be traded.

Commodity Exchange Act: The federal act that provides for federal regulation of futures trading.

Commodity Futures Trading: The federal regulatory agency set up by the government to administer the Commodity Exchange Act which regulates trading on commodities exchanges.

Commodity Pool: An enterprise in which funds contributed by a number of people are combined for the purpose of trading futures or options contracts.

Commodity Pool Operator (CPO): An individual or organization which operates or solicits funds for a commodity pool. Generally registered with the Commodity Futures Trading Commission.

Commodity Trading Advisor (CTA): A person or firm who, for a fee, directly or indirectly advises others as to the value of or advisability of buying or selling futures or options contracts.

Confirmation Statement: A statement sent a Futures Commission Merchant to a customer when a futures or options position has been initiated. The statement shows the number of contracts bought or sold and the prices at which the contracts were bought or sold. Sometimes combined with a Purchase and Sale Statement.

Complainant: The individual that files a complaint seeking reparations against another individual or firm.

Consolidation: A pause during trading activity in which prices move sideways, setting the stage for the next move. During periods of consolidation traders assess their positions.

Contango: A market situation in which prices in succeeding delivery months are progressively higher than in the nearest delivery months; the opposite of Backwardation.

Contract: A term of reference describing a unit of trading for a commodity future or option.

Contract Grades: These are standards or grades of commodities listed in the rules of the exchanges which must be complied with when delivering cash commodities against futures contracts.

Contract Market: A board of trade designated by the Commodity Futures Commission to trade futures or options contracts on a particular commodity. It is commonly used to mean any exchange on which futures are traded.

Contract Month: Month in which a futures contract may be fulfilled by taking or making delivery.

Convergence: The tendency for prices of physical commodities and futures to approach one another, usually during delivery month.

Corner: To secure control of a commodity so that its price could be manipulated.

Correction: A price reaction against a prevailing trend of the market. Sometimes referred to as a retracement.

Cover: To offset a previous futures transaction with an equal and opposite transaction. Short covering is a purchase of futures contracts to cover an earlier sale of an equal number of contracts of the same delivery month.; liquidation is the sale of futures contracts to offset the obligation to take delivery on an equal number of futures contracts of the same delivery month purchased earlier.

Cross-Hedge: Hedging a cash market risk in one futures contract by taking a position in a different, but price-related commodity.

Current Delivery (Month): Also known as spot month. The futures contract will reach maturity and become deliverable during the current delivery month.

Day Order: An order that expires automatically, unless executed, by the end of the trading session on the day it was entered.

Day Trader: A speculator who will initiate and offset a position within a single trading session.

Debit Balance: A condition where the trading losses in a customer's trading account exceed the amount of equity in the account.

Deck: All of the unexpected orders in a floor broker's possession.

Default: The failure to perform on a futures contract as required by the exchange rules, such as failure to meet a margin call.

Deferred Delivery: The distant delivery months in which futures trading is taking place, as distinguished from the nearby futures delivery month.

Delivery: The tender and receipt of an actual commodity or warehouse receipt or other negotiable instrument covering such commodity, in settlement of a futures contract.

Delivery Points: The locations designated by the commodity exchanges at which stocks of a commodity represented by a futures contract may be delivered in fulfillment of the contract.

Delivery Price: The official settlement price of the trading session during which the buyer of a futures contract receives through the clearing house a notice of the seller's intention to deliver at the delivery price.

Delta Value: The expected change in an options price given a one-unit change in the price of the underlying futures contract.

Derivative: A financial instrument, traded on or off the exchange, whose price is directly dependent upon the value of one or more underlying securities, equity indices, debt instruments, commodities, other derivative instruments or any agreed upon pricing index or arrangement.

Disclosure Document: The document provided to and signed by customers which describes fees, risks, etc.

Discount: 1) The amount a price would be reduced to purchase a commodity of lesser grade; 2) Sometimes used to describe the price difference between futures of different delivery months, as in the phrase "July is at a discount to May," indicating that the price of July futures contract is lower than that of May; 3) Applied to cash grain prices that are below the futures price.

Discretionary Account: An arrangement by which the holder of the account gives written power of attorney to another, often a broker, to make buying and selling decisions without notification to the holder; often referred to as Managed Account.

Dual Trading: This occurs when 1) a floor broker executes a customer's order, while at the same time trades for his own account or an account in which he has an interest; 2) Or a Futures Commission Merchant carries customer accounts and also trades, or permits its employees to trade, in accounts in which it has a proprietary interest, also on the same day.

Elasticity: A characteristic of commodities which describes the interaction of supply, demand and price of a commodity. A commodity is said to be elastic in demand when a price change creates an increase or decrease in consumption; the supply of a commodity is said to be elastic when a change in price creates change in the production of the commodity.

Electronic Trading Systems: Systems that allow participating exchanges to list their products for trading after the close of the exchange's open outcry trading hours.

Equity: The dollar value of a futures trading account if all open positions were offset at the going market price.

Exchange: An association of persons engaged in the business of buying and selling commodity futures and options. See Board of Trade and Contract Market.

Exercise: Exercising a call means that you elect to purchase the underlying futures contract at the option strike price. Exercising a put means that you elect to sell the underlying futures contract at the option strike price.

Exercise Price: See Strike Price

Expiration Date: Generally the last day on which an option may be exercised.

Feed Ratios: The variable relationships of the cost of feeding animals to the market weight sales prices, expressed in ratios, such as the hog/corn ratio. These serve as indicators of the profit return or lack of it in feeding animals to make weight.

First Notice Day: The first day on which notice of intent to deliver a commodity in fulfillment of an expiring futures contract can be given to the clearing house by a seller and assigned by the clearing house to a buyer. It varies from contract to contract.

Floor Broker: An individual who executes orders on the trading floor of an exchange for any other person.

Floor Traders: Members of an exchange who are personally present on the trading floor of exchanges to make trades for themselves. Sometimes called Locals.

Forward (Cash) Contract: An agreement where a seller plans to deliver a cash commodity to a buyer sometime in the future. In contrast to futures contracts, the terms in a forward contract are not standardized. Forward contracts are not traded on federally designated exchanges.

Fundamental Analysis: An approach to analysis of futures markets and commodity futures price trends which examines the underlying factors affecting the supply and demand of a commodity.

Futures Commission Merchant (FCM): An individual or organization which solicits or accepts orders to buy or sell futures or options contracts and accepts money or other assets from customers in connection with such orders. Must be registered with the Commodity Futures Trading Commission.

Futures Contract: A legally binding agreement to buy or sell a commodity or financial instrument at a later date. Futures contracts are standardized according to the quality, quantity and delivery time and location for each commodity.

Futures Industry Association (FIA): The national trade association for Futures Commission Merchants.

Gap: A trading day during which the daily price range is completely above or below the previous day's range causing a gap between them. Some traders will then look for a retracement to "fill the gap."

Grantor: A person who sells an option and assumes the obligation but not the right to sell (in the case of a call) or buy (in the case of a put) the underlying futures contract at the exercise price.

Hedging: The practice of offsetting the price risk inherent in any cash market position by taking the opposite position in the futures market. A hedger will use the market to protect his business from adverse price changes.

Inelasticity: A characteristic that describes the interdependence of the supply, demand and price of a commodity. A commodity is inelastic when a price change does not create an increase or decrease in consumption; inelasticity exists when supply and demand are relatively unresponsive to changes in price.

Initial Margin: Customer's funds required at the time a futures position is established, or an option sold, to assure performance of a customer's obligations.

In the Money: An option having intrinsic value. A call is in the money if its strike price is below the current price of the underlying futures contract. A put is in the money if its strike price is above the current price of the underlying futures contract.

Intrinsic Value: The absolute value of the in the money amount; that is, the amount that would be realized if an in the money option were exercised.

Introducing Broker (IB): A firm or individual that solicits and accepts futures orders from customers but does not accept money, securities, or property from the customer. An IB must be registered with the Commodity Futures Trading Commission and must carry its accounts through a Futures Commission Merchant on a fully disclosed basis.

Inverted Market: Futures market in which the nearer months are selling at premiums over the more distant months. Characteristically, this is a market in which supplies are currently in shortage.

Invisible Supply: Uncounted stocks of a commodity in the hands of wholesalers, manufacturers, and producers which cannot be identified properly; the stocks are outside commercial channels but in theory available to the market.

Last Trading Day: The last day on which trading may occur in a given futures or options contract.

Leverage: The ability to control large dollar amounts of a commodity with a comparatively small amount of capital.

Limit : See Position Limit, Price Limit, Variable Limit, and Reporting Limit.

Limit Move; A price that has advanced or declined the limit permitted during one trading session as fixed by the rules of a contract market.

Limit Order: An order in which a customer sets a limit on either price and/or time of execution, as opposed to a market order which implies that the order should be filled at the most favorable price as soon as possible.

Liquidation: To sell (or purchase) futures contracts on the same delivery month purchase (or sold) during an earlier transaction, or make (take) delivery of the cash commodity represented by the futures contract.

Liquidity (Liquid Market): A intensively traded market where buying and selling can be accomplished easily due to the presence of many interested sellers and buyers.

Loan Program: The main means of government agricultural price support operations. The government lends money to farmers at announced rates using the crops as collateral. Default on these loans is the primary method by which the government acquires stock of agricultural commodities.

Long: One who has bought futures contracts or owns a cash commodity.

Long Hedge: Buying futures contracts to protect against possible increased prices of commodities. See also Hedging.

Maintenance Margin: A set minimum margin (per outstanding futures contract) that a customer must maintain. See also Margin.

Margin: An amount of money deposited by both buyers and sellers of futures contracts and by sellers of option contracts to ensure performance of the terms of the contract. Margin in futures is not a down payment, as in securities, but rather a performance bond.

Margin Call: A call from the clearing house to a clearing member, or from a broker or firm to a customer, requiring additional money deposit to bring an account to a certain minimal level.

Mark-to-Market: To debit or credit daily a margin account based on the close of that day's trading session.

Market Order: An order to buy or sell a futures or options contract at whatever price is obtainable when the order reaches the trading floor.

Maximum Price Fluctuation: See Limit Move.

Minimum Price Fluctuation: See Point or Tick.

Momentum Indicator: A line that is plotted to represent the difference between today's price and the price of a given number of days ago. It is measured as the difference between today's price and the current value of a moving average. Often referred to as momentum oscillators.

Moving Average: A mathematical procedure to smooth or eliminate the fluctuations in data. A moving average emphasizes the direction of a trend, confirms trend reversals, smoothes out variations that can confuse the interpretation of the market.

National Association of Futures Trading Advisors (NAFTA): The national trade association of Commodity Pool Operators(CPOs), Commodity Trading Advisors (CTAs), and related industry participants.

National Futures Association (NFA): The industry wide self-regulatory organization of futures trading.

Nearby Delivery Month: The futures contract month closest to expiration.

Net Asset Value: The value of each unit of a commodity pool. It is a calculation of assets minus liabilities plus or minus the value of open positions when marked to the market, divided by the number of units.

Net Performance: An increase or decrease in net asset value exclusive of additions, withdrawals, and redemptions.

Net Position: The difference between the open long (buy) and the open short (sell) contracts held by any one person in any one futures contract month, or in all months combined.

Notice Day: Any day on which notices of intent to deliver on futures contracts may be issued.

Offer: An indication of willingness to sell a futures contract at a given price; the opposite of Bid.

Offset: To take a second futures or options position opposite to the initial or opening position. See also Liquidate.

Omnibus Account : An account carried by one Futures Commission Merchant (FCM) with another FCM in which the transactions of two or more persons are combined and carried in the name of the originating FCM rather than of the individual customers. The identity of the individual accounts is not disclosed.

Open (the): The period at the beginning of the trading session officially designated by the exchange during which all transactions are considered made "at the open."

Open Interest: The sum of all long and short futures contracts in one delivery month or one market, that have been entered into and not liquidated yet by an offsetting transaction, or fulfilled by delivery.

Open Outcry: Method of public auction used for making bids and offers in the trading pits of the commodity exchanges.

Open Trade Equity: The unrealized gain or loss on open positions.

Opening Range: The range of prices at which buy and sell transactions took place during the opening of the market.

Option Contract: A unilateral contract which gives the buyer the right, but not the obligation, to buy or sell a specified quantity of a commodity or a futures contract at a specific price, within a specified period of time, regardless of the market price of the commodity or futures contract. The seller of the option has the obligation to sell the commodity or futures contract or buy it from the option buyer at the exercise price if the option is exercised. See also Call Option and Put Option.

Option Premium: The price a buyer pays for an option. Premiums are arrived at through open competition between buyers and sellers on the trading floor of the exchange.

Open Seller: See Grantor.

Orders: See Market Order, Stop Order, and Limit Order,

Original Margin: See Initial Margin

Out of the Money: A call option with a strike price higher or a put option with a strike price lower than the current market value of the underlying asset.

Overbought: A technical opinion that the market price has risen too steeply and too rapidly in relation to underlying fundamental factors.

Oversold: A technical opinion that the market price has declined too steeply or too rapidly in relation to underlying fundamental factors.

Par: A particular price 100 % of principle value.

Parity: A theoretically equal relationship between farm product prices and all other prices.

Pit: The area on the trading floor of some exchanges where trading of futures and options contracts is conducted by open outcry.

Point: The minimum fluctuation in futures prices or options premiums.

Pool: See Commodity Pool.

Position: A commitment, either long or short, in the market.

Position Limit: The maximum number of futures contracts that one can hold in certain regulated commodities, as determined by the Commodity Futures Trading Commission.

Position Trader: A trader who either buys or sells contracts and holds them for an extended period, as distinguished from a day trader, who will normally initiate and liquidate a futures position during the course of a single trading day.

Premium: Refers to 1) the amount a price would need to be increased in order to purchase a better quality commodity; or 2) a future delivery month selling at a higher price than another; or 3) cash prices that are above the futures prices.

Price Discovery: The process of determining the price levels of a commodity based on supply and demand factors.

Price Limit: The maximum price advance or decline from the previous day's settlement permitted for a futures contract in one trading session.

Primary Markets: The principle markets for the purchase and sale of a cash commodity.

Principal: Refers to 1) a sole proprietor, general partner, officer or director, or person of similar status, who has the power of influence over the activities of the entity; or 2) any holder or any beneficial owner of 10% or more of the outstanding shares of any class of stock of the entity; or 3) any person who has contributed 10% or more of the capital of the entity.

Public Elevators: Grain storage facilities, licensed and regulated by state and federal agencies, in which space is rented out. Some are also approved by the commodity exchanges for delivery of commodities against futures contracts.

Purchase and Sale Statement (P&S): A statement sent by a Futures Commission Merchant to a customer when a futures or options position has been liquidated or offset. The statement contains all of the transaction information.

Put (Option): An option that gives the option buyer the right but not the obligation to sell the underlying futures contract at a particular price on or before a particular date.

Pyramiding: The use of unrealized profits on existing futures positions as margins to increase the size of the position, normally in successively smaller increments. For instance, the use of profits on the purchase of five futures contracts as margin to purchase an additional four contracts, whose profits will in turn be used to margin an additional three contracts.

Quotations: The actual price or the bid or ask price of either cash commodities or futures or options contracts at a particular time.

Range: The difference between the high and low price of a commodity during a given trading session, week, month, year, etc.

Reaction: A short term countertrend movement of prices.

Recovery: An upward movement of prices following a decline.

Registered Commodity Representative (RCR): Seen Broker.

Reparations: Compensation payable to a wronged party in a futures or options transaction.

Reparations Award: The monetary damages a respondent may be ordered to pay a complainant.

Reporting Limit: Sizes of positions set by the exchange and/or the Commodity Futures Trading Commission at or above which commodity traders must make daily reports to the exchange in terms of all the details of the transaction.

Resistance: The price level where a trend stalls. It is the opposite of a support level. Prices must build momentum to move pass a resistance.

Respondents: The individual or firm against which a complaint has been filed and a reparations award is being sought.

Retracement: A price movement in the opposite direction of the prevailing trend. See also Correction.

Round Lot: A quantity of a commodity equal in size to the corresponding futures contract for the commodity, as distinguished from a job lot, which may be larger or smaller than the contract.

Round Turn: A completed futures transaction involving both a purchase and a liquidating sale, or a sale followed by a covering purchase.

Rules (NFA): The standards and requirements to which participants who are required to be Members of National Futures Association must subscribe and conform.

Sample Grade: The lowest quality acceptable of a commodity for delivery in satisfaction of futures contracts. See Contract Grades.

Scalper: A trader who trades for small, short-term profits during the course of a trading session, rarely carrying a position overnight.

Security Deposit: See Margin.

Segregated Account; A special account used to hold and separate customers' assets from those of the broker of firm.

Selling Hedge: See Short Hedge

Settlement Price: The daily price at which the clearing house settles all accounts between clearing members for each contract. This is the closing price, which is used as the official price in determining net gains or losses at the close of each trading session.

Short: One who has sold futures contracts or the cash commodity.

Short Hedge: Selling futures contracts to protect against possible declining prices of commodities. See also Hedging.

Speculator: One who tries to profit from buying and selling futures and options contracts by anticipating future price direction.

Spot: Usually refers to a cash market price for a physical commodity that is available for immediate delivery.

Spread: The simultaneous buying and selling of two related markets in the expectation that a profit will be made when the position is offset.

Stop Loss: A risk management technique used to close out a losing position at a given point. See Stop Order.

Stop Order: An order that becomes a market order when a particular price level is reached. A sell stop is placed below the market, a buy stop is placed above the market. Sometimes referred to as a Stop Loss Order.

Strike Price: The price at which the buyer of a call (put) option may choose to exercise his right to purchase (sell) the underlying futures contract. Also called Exercise Price.

Support A: price level at which a declining market has stopped falling. It is the opposite of a resistance price range. Once this level is reached, the market trades sideways for a period of time.

Technical Analysis: An approach to trading that is based on the assessment and examination of patterns of price change, rates of change and changes in volume of trading, open interest and other statistical indicators. See data is usually charted.

Tick: The smallest allowable increment of price movement for a contract. Also referred to as Minimum Price Fluctuation.

Time Value: Any amount where the option premium exceeds the option's intrinsic value.

Traders: Generally people who trade for their own account or employees or institutions who trade for their employer's account.

Uncovered Option: A short call or put option position which is not covered by the purchase or sale of the underlying futures contract or physical commodity.

Underlying Futures Contract: The specific futures contract that the option conveys the right to buy (in case of a call) or sell (in case of a put).

Variable Limit: A price system that allows for larger than normal allowable price movements under certain conditions. In periods of extreme volatility, some exchanges permit trading at price levels that exceed regular daily limits.

Variation Margin Call: Additional margin deposited by clearing member firm to an exchange clearing house during periods of great market volatility or in the case of high-risk accounts.

Volatility: A measurement of the change in price over a given time frame.

Volume of Trade: Number of contracts traded over a specified period of time.

About the Author

With over two decades of experience as an Entrepreneur and Education Consultant, I have had the privilege of aiding clients and students across various specialized domains, including finance, business management, real estate, and federal and state tax law. My deep-seated passion lies in the realm of teaching, where I relish the opportunity to impart practical and invaluable knowledge to eager learners, opening doors to fresh concepts and innovative ideas.

With this book my mission has been clear and concise: to demystify complex subjects like futures trading by crafting a user-friendly, step-by-step tutorial. While this tutorial may come across as straightforward and uncomplicated, it is, in fact, a comprehensive and in-depth exploration of the subject matter.

If you've enjoyed this book and have gained value from it, please consider posting a review on Amazon. I'm always happy to hear your comments and suggestions.

Click here or scan the QR code below to leave your honest review.

SCAN HERE!

References

Chicago Board of Trade Commodity Trading Manual, (revised periodically) Board of Trade of the City of Chicago, IL.

Elliot, R. N., The Wave Principle, (1938) Elliot, New York, NY.

Eng, William F., The Day Trader's Manual: Theory, Art and Science of Profitable Short-Term Investing, (1993) John Wiley & Sons, Inc., New York, NY.

Kaufman, Perry J., Handbook of Futures Markets: Commodity, Financial, Stock Index, and Options, (1984) John Wiley & Sons, Inc., New York, NY.

Stevenson. John The Art of Options Trading (2016) New York, NY

Kolb, Robert W., Understanding Futures Markets, 3rd edition (1991) Kolb Publishing Inc., Miami, FL.

Murphy, John J., Technical Analysis of the Futures Markets: A Comprehensive Guide to Trading Methods and

Applications, (1986) New York Institute of Finance, New York, NY.

Niemira, Michael P., and Gerald F. Zukowski, Trading the Fundamentals: The Trader's Complete Guide to Interpreting Economic Indicators and Monetary Policies, (1994) Probus Publishing Company, Inc., Chicago, IL.

Rothstein, Nancy H. and James M. Little, The Handbook of Financial Futures: A Guide for Investors and Professional Financial Managers, (1984) McGraw-Hill, Summit, PA.

Wiest, Robert E., You Can't Lose Trading Commodities, Beacon Publishing, Naples, FL.